Simplicity and Complexity in Games of the Intellect

Lawrence B. Slobodkin

HARVARD UNIVERSITY PRESS
Cambridge, Massachusetts
London, England

Copyright © 1992 by Lawrence B. Slobodkin
All rights reserved
Printed in the United States of America
10 9 8 7 6 5 4 3 2

First Harvard University Press paperback edition, 1993

Library of Congress Cataloging-in-Publication Data

Slobodkin, Lawrence B.
Simplicity and complexity in games of the intellect / Lawrence Slobodkin.
p.   cm.
Includes bibliographical references and index.
ISBN 0-674-80825-8 (cloth)
ISBN 0-674-80826-6 (pbk.)
1. Science—Philosophy.   2. Science—Methodology.   3. Simplicity
(Philosophy)   4. Complexity (Philosophy)   I. Title.
Q175.S578   1992
501—dc20   91-2372
CIP

Excerpt from "East Coker" in FOUR QUARTETS, copyright 1943 by T. S. Eliot and renewed 1971 by Esme Valerie Eliot, reprinted by permission of Harcourt Brace Jovanovich, Inc.

Contents

This book was written to be read by Yan and Mathew
when they have really learned to read.
It is dedicated to the memory of
G. Evelyn Hutchinson (1903–1991),
who taught me to really read.

The Opening

We each see the world through very personal perceptual spectacles. It is obvious to observant adults that two-to-four-year-olds develop a personal set of natural laws—not necessarily correct, nor intellectually deep, but nevertheless adequate summaries of their perceptions. Things that smoke are hot, kittens can scratch, parents can grow angry under suitable stimuli, and so on. A child sorts natural phenomena into classes of events that are expected to have different properties and therefore must be approached differently. Scientists and philosophers use a finer and more self-conscious system of categorizations. Sometimes the children are wrong in their expectations of how things will function. But sometimes older intellectuals are also wrong.

The natural laws inferred by small children, or "discovered" by scientists, and also representations of the world by artists and authors, are of necessity either simplifications of the bewildering collection of possible observations and speculations about nature, or else they are complications of simple aspects of nature whose simplicity is not apparent. The intellectual roles of simplification, complication, and their extreme manifestations as minimalization and elaboration are my theme. Understanding simplicity is not always easy because a cluster of words and meanings must be considered in many different contexts.

At least four different ideas related to simplicity lie close together. The distinctions among "simple," "simplified," "simplistic," and "minimalistic" are perhaps made clear by considering their respective opposites: "Complex" or maybe "difficult" is the opposite of "simple," "complicated" of "simplified," "obfuscated" of "simplistic," and "ornate" or "elaborate" of "minimalistic." More or different

1

subdivisions of the idea could be defended, but these will do for a start.

The Simple and the Complex

For both objects and concepts the most obvious meanings of "simple" relate to homogeneity in structure and the absence, or at least small number, of discernible parts or partitions. Sometimes "simple" means that a phenomenon has clear underlying causes. A simple dress, painting, symphony, theory, or society has fewer lines, curves, colors, classes, or other internal differentiations than a more complicated one. A simple theory has fewer terms than a more complex one.

Most natural objects can be considered simple when examined on an appropriate scale. A smooth, unpainted, metal ball bearing can be described by a relatively small number of assertions. After defining it as a sphere and giving its diameter, weight, chemical constitution, and hardness, not much is left to say. A glass marble can also be described simply, as a multicolored glass sphere with a swirled color pattern. However, the complexity of the swirl patterns in the marbles are very highly appreciated, on an esthetic level, by the children that play with them, and in fact the swirls cannot be completely described except by using advanced mathematics.

Most description involves simplification. In some esthetic and intellectual contexts simplicity is always to be preferred, other things being equal. Scientists in particular prefer quantitative theories that use the smallest number of terms, the lowest number of significant figures, and the lowest exponents, while still meeting agreed criteria for descriptive accuracy.

In science the deepest form of simplicity is called "elegance." Elegance is the primary touchstone for the most advanced of theories. At this level, simple-looking, nonintuitive mathematical or logical formalisms predict or explain universal aspects of nature. Some of these formulations require profound intellectual effort to understand and enormous research budgets to empirically test. Sometimes, even to conceive of empirical tests is beyond the limits of present intellectual accomplishment.

The opposite of simple is "complex," and some things really do seem complex, requiring many parts and many differentiations in order to accomplish a particular job. The array of belts, wheels, and pulleys in a nineteenth-century factory all had to be there to transmit power from a central steam engine or water wheel to the correct places. The maze of sheets, stays, halyards, and sails on large square-rigged sailing ships is another example of required complexity. Of course, finding ways to eliminate this complexity is all to the good. Fore-and-aft rigging simplified sailing, and this was important progress. (It had the misfortune to arrive at around the same time as the steamboat, so that elegant sailing was invented just in time to become obsolete.)

Most tools are beautifully and necessarily simple. Hammers, crowbars, and chisels have shapes that do not admit of much tampering. A boomerang of a shape that is either simpler or more complex than usual will not return. But notice that the boomerang can be painted with the most complex designs imaginable, indicating that esthetic complexity is compatible with functional simplicity.

Some jobs cannot be made simpler, and in these cases there is no discredit to using a complicated procedure for getting them done. However, reasonable people generally disapprove of apparently gratuitous complexity in performing what seem to be relatively simple tasks. This is exemplified by the popular response to the legal profession. Lawyers are accused of creating complications to enhance their incomes. Lawyers defend themselves by saying that without their efforts all manner of business would become even more seriously complicated. Both lawyers and their critics agree that unnecessary complication is not good.

Even if complicated objects can only be partially described, many of them can be considered as consequences of some system of natural laws, which constitutes a kind of complete description. The surface of a wavy sea may be impossible to describe, but some aspects of the pattern of waves may be inferred from explicit mathematical formulations. Also, on the scale of microscopic crystal structure, the apparently simple ball bearing is as complicated as any irregular lump of metal.

The classical physical laws of the lever, screw, inclined plane, combined with the most elementary concepts of gravitation and hydrodynamics, are all that are involved—aside from talent and ingenuity—to design even the most complicated water-powered factory. Elementary principles of metallurgy, wood, and fiber craft, combined with manual skill, can make those designs into reality. Description of the laws underlying a particular factory requires less specification than describing the factory itself. In this sense, the laws of science are often simpler than the systems that use them for technological ends. Similarly, the relatively simple-sounding assertions of principle found in the United States Constitution underlie the apparent tangles of everyday law.

At one extreme, then, there exist simple objects and systems, which serve some function or fill some natural role with a very small number of internally distinguishable states, parts, or conditions. At the opposite extreme are complex objects and systems that cannot be reduced to simple ones. Natural laws can also be ordered in their simplicity, but not as cleanly. Newtonian gravitation seems simpler than the still elusive laws of biological development, until we recall the deep problems in explaining gravity itself.

The Simplified and The Complicated

Intellectual work is performed in a simplified arena. Sometimes the simplification is imposed by the task or the tools at hand; sometimes the process of simplification must be imposed on a world which does not provide any good hints as to the most effective simplification. A musical instrument only produces sound when handled in a particular way, and in that sense the interaction between the player and the instrument has been simplified. A chess game only admits moves made according to a simple set of rules. The sequence of notes or moves played may demonstrate genius or virtuosity. On the other hand, in science or literature there are almost no general simplifying restrictions, and being able to simplify effectively is the mark of a virtuoso.

Perhaps child prodigies can occur only if the intellectual arena imposes its own simplifications. For example, Mozart composed harpsichord pieces at the age of four, and many children can play simple tunes at the age of four or five. There has never been a scientific prodigy to equal that precocity. Darwin the child loved plants and animals but was not a prodigy.

Four-year-olds may become chess or musical prodigies because the worlds of chess and music are presented to the child in simplified form. To be a prodigy in science would require the child to make the simplification for himself. Perhaps one reason that four-year-olds are less accomplished scientists than some PhDs is that the four-year-olds accept the world whole, except in special game-like contexts, while the training of PhDs teaches them how to ignore large segments of reality. Most natural systems are very difficult to understand or communicate about. However, the natural world can be deliberately "simplified" before being described. It is possible, and sometimes appropriate, to replace actual systems with simulations or models. These then become the objects of study or manipulation.

Excessive complexity may be disturbing in ways that are much more immediate than the construction of scientific theories. Archaeologists and anthropologists have recently become concerned with the development of complex societies. They accept a definition of complexity which includes the number of social subclasses that can be found. A society with an aristocracy is more complex than an egalitarian society. One of the suggested explanations for the origin of complex society is that if a lot of people must interact with one another on a personal level, it seems to get on their nerves. Egalitarian groups of more than around thirty or forty people are near the limit of the individual's tolerance of complexity. This excess of interactions is simplified for the individual by the establishment of hierarchical complication. If a chief tells everyone what to do, then only one set of orders need be heard rather than the voices of tens of others. Complex societies may develop, therefore, to serve the function of simplifying the immediate world of their inhabitants.

It is tempting to believe that perhaps complementarity between simplicity on one level and complexity on another always exists. For example, if verb endings disappear in the evolution of a language, as in English, pronouns and auxiliary verbs may become more elaborate. Religious doctrines that have complex legalistic systems may tend toward simple theologies, and the converse. On the level of personal psychology, individuals tend to build personal simplifications of their ambient world, perhaps because there is a limit to acceptable cognitive dissonance. To simplify a problem to the point that it can be lived with may be almost as good as solving it.

Returning to the context of scientific analysis, if a system shows complicated changes in time, one can still simplify it by examining it for only a short interval, a very narrow time slice, during which most changes would be negligible. Similarly, spatial complication can be avoided by taking a very small region. Even a very complex phenomenon will appear simple on a small enough scale. For example, if one is confronted with a system describable by a set of curves, one can choose a very small segment of the curve for examination. As the extent of the curve examined becomes smaller and smaller, the apparent curvature usually becomes less and less, until the bit of curve being examined can be considered a straight line. This is often helpful because equations for straight lines are of the form $y=ax+b$. If the constants a and b have only a reasonable number of significant figures, then the equations for straight lines are simpler than those for curves. However, putting the little pieces of well-described linear world together to regain the whole curve may pose problems and may not be easier than dealing with the entire system at once.

The new and very stylish mathematical field of fractal geometry focuses on situations for which curves do not approach straight lines no matter how small the region of focus. The complexity of curvature of a seacoast seems to stay the same on a scale of miles, feet, or inches. But of course fractal geometry is hailed as a simple way of thinking about this irreducible complexity![1]

Sometimes the best approach is to deal with an entire system, with all of its extension in time and space, by deliberately simplifying our conception of it all at once. A common procedure is to use the assumption that "all other things are and remain equal." If the entire universe outside the particular system I want to describe, and all the properties of that system other than the ones that interest me, were to remain constant, then I could perhaps construct a reasonably simple description of limited scope. This kind of hopeful simplification of a large problem is a common intellectual strategy in the less tractable sciences. Just as in trying to put the small pieces of a problem back together to make a whole is difficult, releasing the *ceteris paribus* condition may not be easy. Perhaps the complexity that has been discarded will become important.

The opposite of simplification is "complication," adding new parts or concepts to a simpler system. This may be done for several reasons. For example, compare the dozens of dials, buttons, and switches in a modern airplane cockpit with the small number of controls in a World War I fighter plane, but then compare the speed, reliability, and safety of modern planes with those of sixty years ago. Complicating the cockpit was a trivial price to pay for the improvement. When the problem of designing reasonably safe and reliable automobiles had been solved, then radios, cigarette lighters, telephones, and even bars, toilets, beds, and television sets were added. From the simple functional cars and trucks finally emerged land cruisers and mobile homes. As transportation, nothing had been gained, but the role of the automotive machine in its owner's system of values and desires had been considerably expanded.

A simplified theory may be made to more closely approximate reality by permitting complexity to return in a stepwise fashion. The simplified mathematical models that were originally designed for discussions of the growth of single species of organisms in constant environments had a total of two constants. These were then complicated by later investigators to deal with ideas about predation, multispecies interaction, symbiosis, and so on. Each of these ramifications added at least one more constant and occasionally whole functions. The theory that began as a simplification was eventually much too complicated to be of interest to anyone other than its creators and their students. At a certain point excess complication becomes useless.

Not all complication is designed to be useful, nor is all simplification in the interest of solving problems. The most common role for both simplification and complication is esthetic. Representational and abstract graphic art, music, cooking, dress, and almost all human endeavors that are even partially esthetically motivated can be simplifications or complications of some standard approach. Even fundamentally useless complications of actual machines can have esthetic value, as demonstrated by automobiles' chrome and tail fins and by the ways commercial automobile designers and customizers treat the body of a car as a piece of sculpture. The fictional machines filmed in Chaplin's "Modern Times" or Lang's "Metropolis" have extra useless moving parts for esthetic reasons. Extra rope is added as needed for dramatic entrances to the sailboats in pirate movies.

Nonpragmatic complication will be considered further in the context of "obfuscation" and "elaboration."

The Simplistic and the Obfuscated

Simplification and complication are neutral activities, whose value is to be judged case by case. The decision to simplify a system by choosing to ignore some of its complexities is an easy one to make, but its proper execution requires a happy combination of talent, insight, background information, and intelligence. Some simplifications and complications are brilliant. An unskilled, failed attempt at simplification falls into the category of the "simplistic."

Similarly, adding complexity to either the perceived world or a simplification of it must be done carefully and with a discrete purpose, lest complication becomes "obfuscation." In creating a simplified or complicated system, the divergence from reality is acknowledged. To be simplistic or to obfuscate betrays misunderstanding, or worse, an attempt to hoodwink.

As a class of examples of both the simplistic and the obfuscatory, consider the more popular pseudosciences, which attempt to satisfy the hunger to predict our personal futures or at least provide foci of blame for our own inadequacies. Graphology, phrenology, Sunday supplement personality quizzes, pop theories of evolution, vulgar Marxism, racial and ethnic bigotry, and astrology all share gratuitous simplifications and complications. Often these involve reification of parts as wholes, plausible but nonoperational causes, untestable conclusions, and emotional appeals all wrapped in obscurantic clouds of words, charts, and diagrams, or even in equations and computer programs.

Astrology and cheiromancy, the reading of hands, may be the most venerable of the extant pseudosciences, but astrology has a more highly organized pseudo-theoretical rationale. Namely, it is based on real measurements of astronomical events, time, and location. In fact, seasonal shifts in the apparent location of the stars, and the movements of the sun, moon, and planets against the background of stars, are the only really regular, predictable, and readily observable natural phenomena in existence. The regularity of the stars, sun, and planets proclaims the regularity and the lawfulness of the world. The angle of the sun is so low at the winter solstice that a sunbeam can climb higher up a wall, or extend deeper into an appropriately oriented tunnel, at noon on that date than any other time of year. Holes for sunlight and calibrations on the floors and walls of many Italian cathedrals, the megalithic tomb at Newgrange in Ireland, the large stone circle at Stonehenge, and prehistoric arrangements of rocks in New Mexico all take advantage of this fact to reset their calendar. Tides, best planting dates, eclipses all follow the regularity of the planets and stars. Plants and animals regulate their lives by the calendar.

Birds build nests, then lay eggs, then hatch the eggs, and only then feed their young. How can a bird know when to build its nest so that the insects will arrive on time to feed the eventual young? Sometimes spring warms early, sometimes late, but for any given locality there is an average date for the appearance of insects of the appropriate kind and abundance. The best predictor of the arrival of insects is the calendar, and the most available calendar for organisms is the day length combined with the sign of day-length change. North of the equator, if the days are lengthening, it is between December 22 and June 22. Exactly where in this half year one is located is given by the day length at any given latitude. Swelling of the testes of male birds and ovarian growth of females are prerequisite to courtship and nest building and to actual reproductive activities such as fertilization and egg laying. These are coordinated by changes in day length.

If one believes in causality and is confronted with the need to predict the future, it seems obvious that one should proceed from the more predictable to the less predictable features of the world. If even the lives of plants and animals follow the stars, what could be more logical than assuming that the lives of humans also must follow the stars? To each moment in time there corresponds a particular configuration of sun, moon, stars, and planets which will not recur for many centuries. It seems eminently reasonable, at a superficial glance, that the precise time and date of birth, or perhaps of conception, should have some influence on future properties.

Modern astrologers use the best astronomical data and include as many heavenly bodies as possible at the time of the birth. To the degree that the precise time and location is not known, even an approximation will supposedly do, as witnessed by newspaper horoscopes that apply to individuals born within approximately one lunar month. Of course astrologers concede that "accuracy" is thereby sacrificed. For example, the seventeenth-century biographer John Aubrey would begin his studies of poets and other intellectuals by determining as precisely as he could their birth times and then attempting to correlate properties of their personal lives or their work with the position of the stars and planets at birth. He concluded that he really could not find any correlations, but he attributed this to his inability to collect really precise birth-time data.[2]

As of today, no correlations of the sort Aubrey sought have been found. The theory of horoscopes can be made as complex as one pleases and may even use computers to keep all the heavenly trajectories straight. Yet its validity is still no better than the simplistic assumptions with which it began. Of course, this little antiastrology sermon is being preached to the converted; most literate readers disdain astrology already. But if scientific analysis of difficult and complex subject matter must rely, in part, on simplification, and if no uniform, testable theory of how this simplification can be done exists, then might we not conclude with near certainty, that many respectable-sounding intellectual schemes may be no more valid than astrology?

Some people have suggested that Freudian and Jungian psychoanalysis, and related psychoanalytic schemes, are in this category. Grunbaum asks what the real evidence is for the reality of the Ego, Id, Superego, engrams, or racial memories?[3] The counterargument is that Grunbaum and others misunderstand the situation by failing to enter its "hermeneutic circle." For the moment a hermeneutic circle can be thought of as the complete logic of the system, which can only be appreciated by those who suspend disbelief long enough to see all the system's parts in context. Unfortunately, astrology also has a hermeneutics.

The Minimal and the Ornate

Minimalism is one of the foci of this book. In this introduction I will supply only sufficient examples to act as a simple but not minimalistic definition. At one level, minimalism is an extreme form of simplification, in which style is at least as important as substance. Minimalist approaches step back from the usual way of doing something, or of seeing a subject, and thereby illuminate that way and its context. Minimalistic bathing suits are now clichés, but the first bikinis were comments on modesty and prurience and on the social and erotic ramifications of the sport of swimming.

Visual arts most obviously and repeatedly spawn minimalist movements, beginning with the faceless heavy-hipped paleolithic goddesses and extending to the sad horizontal stripes by Mark Rothko. But minimalism also characterizes certain intervals or schools within all fields of intellectual work. Ultimate minimalism may be pure commentary, without anything of what is usually considered content. John Cage's piano composition 4:33, which consists of a pianist in a concert hall in formal dress sitting in silence at the piano, is a statement about music and its role, as well as about the relation between performer, instruments, and audience, and perhaps other things. The 120 bricks which Carl André stole from a building site and then arranged in a rectangle on the floor of a London museum, after they had been purchased for $12,000, took on artistic meaning as art commenting on art, by virtue of their price, provenance, and presence on the museum floor. Similar comments apply to almost blank canvasses and deliberately plotless novels and films.

At its best, minimalistic art is effective because both the artist and audience share a rich, if not necessarily deep, understanding of the current state of the field. The audience completes the work for themselves, perhaps with radical new insights. Something is gained by knowing that one has rather enjoyed "silence" in a concert hall. Specifically, an audience sitting in a quiet concert hall listens intensely to ambient sounds that are usually ignored.

It is also possible to comment on intellectual problems by producing gratuitous complexity for its own sake. To be deliberately ornate captures most of the idea of the opposite of the minimal. The coiffures of the ladies of the eighteenth-century French court, the novel *Tristram Shandy*, Jackson Pollock's later paintings, numerological speculations of the Jain sect, all evidence delight in ornateness.

What Is the Rest of This Book About?

So far, I have demonstrated that a constellation of seriously different meanings cluster around the vague notion of simplicity. These are important in many different intellectual, esthetic, and practical contexts. The following chapters will explore some of these. But my inquiry cannot stay within the safety of conventional disciplinary borders. The discussion will be neither properly erudite nor turgidly philosophical. My motive is to explore the general premise that simplicity, its associated concepts, and their role deserve more attention for at least two reasons. First, as a possible danger, lest they contribute to illusion and—worse—to delusion. Second, as an inescapable part of intellectual work of all kinds, and of attempts to deal with difficult practical problems.

My concern for this problem grew out of studying biology. Biologists reiterate how complex organisms and ecological systems are, often in the context of excusing the relatively slow rate of intellectual advances in some investigative areas. Partially to avoid problems of complexity, for several decades I have studied what are acknowledged to be the simplest of multicellular animals, the polyps, often in simplified laboratory systems. I eventually realized that I was not at all certain what was meant by "simplicity." This book grew out of that uncertainty.

The sequence of the initial chapter drafts did not correspond to the order in which chapters appear in the text, but as work proceeded, an inevitable order seemed to arise. It may be helpful to describe the logic of that sequence, and in the process to provide a brief outline of the contents of each chapter.

Surveying the intellectual role of simplicity and simplification seemed to require some exploration of art, religion, and science, since there is very little intellectual work which cannot be assigned to one of these three categories. I began by assuming that art is a matter of taste, religion a matter of belief and opinion, and science a matter of fact, and that the meanings and uses of simplicity would show clear differences in the three areas. I thought that the book would therefore divide into three sections: art, religion, and science. These neat distinctions did not hold. For example, art usually involves much more than technique and taste. Religion permeates, logical properties that generate the patterns of simplicity and complexity in the art of dining.

When I look at other arts, in Chapter 5, these patterns are seen to be complicated by issues of money, morality, and politics. The bulk of this chapter is devoted to defining art itself in the simplest way possible and then tracing one narrow thread of art history which leads from representational and narrative painting through to minimalism. This transition illustrates again roles for simplicity and complexity in matters of taste, but in a context of art as commodity and social commentary as well as esthetic pursuit.

For millennia, religion was a main path for approaching reality, but centuries ago science began to partially replace this role of religion. The role of ideas of simplicity in science are considered in Chapters 6, 7, and 8. The first of these deals with the historical beginnings of modern science and with the esthetic and emotional appeal that science seems to have for young students. The doing of science provides a source of satisfying simplicity and certainty. This may involve an elegant hypothetical deductive system, generating the apparent world from not obviously apparent assumptions, as in the Newtonian revolution discussed in Chapter 7, or it may involve deductions of nonobvious processes from examination of a large number of everyday facts, as in the Darwinian revolution discussed in Chapter 8. As in art, a strong esthetic component can be found in the simplifications and minimalizations that enter into the construction of science, but, as in religion, a conviction of dealing with reality lends seriousness and irreversibility to the history of science.

Just as art and religion interact, so that, for example, much of the best of the world's literature lies between art and religion, so the accomplishments of early science lie on the borders of religion. Also art and science interact. In the most exciting parts of science, esthetic criteria of simplicity and minimalism are certainly as strong as in the most frivolous of the arts. But this is not at all surprising since I am treating art and science within the context of a great intellectual game in which simplicity is a valued playing style.

In Chapter 9 I consider a different way that some people deal with simplicity and complexity—namely, by assuming an identity, or at least a general correlation, between simplicity, virtue, and morality on one side and between complexity and licentious immorality on the other. This takes us out of the realm of intellectual games.

Having at least touched on as many aspects of simplicity and complexity as I could easily understand, in Chapter 10 I turn to the philosophical and mathematical analysis of the subject. In this chapter I was looking for a satisfying synthesis, but I was in part disappointed. The philosophers either reenter the intellectual game, or, if they reject such frivolity, do not seem particularly comprehensible. Mathematics is somewhat more satisfying, since the consequences of various assumptions about complexity are more formally determinable. In computer science, simplicity and complexity assume their most tangible meaning and greatest practical significance.

I enjoyed writing this book, but now that time has come for others to read it I feel personally exposed, no longer protected by the conventions of scientific prose. Scientific exposition assumes neatly packaged subject matter and requires that writer and reader trudge through required dull bits, hoping to be rewarded by those parts that are more fun. This is not that kind of book, nor is it written in the mode of a morally righteous pedagogue blessed with a sudden enlightenment. It is an account of an intellectual sailboat voyage or walking tour. Those modes of travel are not the quickest ways to reach a destination, but if the voyage itself is the destination, if the travelers are healthy and the weather fair, there are no better ways to enjoy the parts of the world that are invisible from an airplane or motorcar. One learns intimate things about oneself and the world by walking or sailing. I hope you have as much pleasure in the reading as I had in the writing.

senior and high-status member of the troop, made a new discovery. She took a handful of sandy wheat to the seashore, tossed it in and scooped up the floating grains of wheat. This provided her with food that was moistened, seasoned, and sand-free. The procedure caught on almost immediately.

When macaque mothers, carrying their young on their backs or against their bellies, washed wheat in this way, the baby monkeys were gently dipped in the sea. When these young monkeys became independent, they were able to swim. Swimming had never before been observed in the troop. The young swimmers then discovered a different way of gathering food—theft. They would suddenly splash the older animals while they were washing their wheat. When the older animals retreated the young ones rushed in, scooped up the wheat, and swam to the safety of the offshore island.

I have described events in what I consider to be the cultural history of a group of monkeys. Such changes are characterized by some particular animal inventing a new behavior which materially changes how it relates to its particular environment. This innovation is then transmitted between individuals and down the generations, without any genetic changes having occurred in the population.

Even in the telling, the focus of the description of the landscape had to change. The offshore island, an inconsequential detail at the start of the description, became important at the finale. It was not the investigators that assigned differential importance to different aspects of the landscape, but rather that the animals made the assignment in the context of their behavior. These monkeys have a very different relation to their environment than do the other troops of macaques in Japan. Each troop has its own behavioral patterns. In the Japanese mountains, the members of one troop spend a large share of their time sitting in volcanic hot springs to stay warm during the snowy winters.

Had there been no observers present during these cultural changes, we might have attributed the difference between the swimming, stealing, food-washing monkeys and the Jacuzzi-sitting monkeys to genetic differences. As it is, we must concede a tremendous flexibility in behavior and a major transmission of behavior by nongenetic means in these monkeys.

Cultural transmission, in this limited sense, is not confined to primates or even to mammals. At least one case well documents nongenetic transmission of behavior in birds.[9] In England in the early fifties a new type of milk bottle was left on doorsteps. Within months an epidemic of cream thefts spread from three foci and threatened to cover the island. The thieves were chickadee-like birds, the great tits. At least three individual birds, independently and in different parts of the county, had discovered how to peck through the new bottle lids. Their flock mates had watched them and learned from them, and top cream was soon unsafe in all of England, until the bottle lids were redesigned once again. (Homogenized milk, and the end of home delivery, changed the parameters of the problem.)

It may be objected that I have stretched the idea of culture. What is missing in the cultural history of macaques is any sense that the innovation was necessarily "thought out." Perhaps the first washing of the sandy sweet potatoes was an accidental consequence of the monkey taking a drink with a sweet potato in her hand? Also anthropologists usually associate culture with language and with artifacts that persist from generation to generation.[10] I do not feel strongly on the point. What is important is that not only is the world of organisms simplified by the nature of their nervous system but that, at least for some organisms, the way in which it is simplified need not stay constant, even in the absence of evolutionary or genetic change, and also that innovative behavior can be transmitted among organisms and from generation to generation, even in the absence of language. Neurological and behavioral mechanisms permit many organisms to build a sensory world, structured to meet their needs. In that sense simplification has occurred. Since introspective evidence and information about very young children, nonhuman primates, carnivores, and birds, as well as facts of neuroanatomy and neurophysiology, all point in the same direction, I claim that our sense of reality itself involves simplification, even before conscious thought is applied.

Introspective Normative Self-Image of Primates

When we refer to minimalism or simplification in the development of religion, art, or science we mean something much more deliberate than a Gestalt perception of an oscillating cube. To actively simplify is to deliberately construct a theory or object that purports to represent a simplified reality, at least for the time being or until some purpose has been served. This requires the capacity to manip-ulate, and even play with, concepts. To develop a conscious theory is to modify a preexisting conception of the world.

How can we tell a person's ideas about the world? When dealing with most humans we can ask them. If we get no reply at all, it may be that we are either using different languages or that for at least one of us language use or comprehension is imperfect. A serious problem for pediatricians is that many of their patients cannot speak any language comprehensibly. But even when children cannot de-scribe their own anatomy in words, they may be able to respond to the request, "Show me where it hurts!" by pointing. That is, some kind of self-reference is possible before speaking ability.

Parents also know that babies have plans, goals, and opinions before they can speak. A baby has a sense of self before it has the words to discuss the subject. Merleau-Ponty attached great impor-tance to experience with mirrors in both the development of a sense of self in babies and in determining whether or not there exists a sense of self. The response of a two-year-old to a mirror leaves no question that the baby has a sense of curiosity about itself.[11]

Animals from fish to man will show some response to images in mirrors. For example, all aquarists know that Siamese fighting fish expand their fins, flush their colors, and perform elaborate fighting movements if provided with mirrors. I have even seen some do it so well that they defeat themselves and hide from the mirror. Gen-erally, the responses of animals to mirrors are as though they were seeing some other animal of their own species; except for people and some primates, there is no indication that animals have any sense of seeing themselves.

Great apes will try on hats or put bowls on their head in front of a mirror or use mirrors for intimate examination of themselves, much as humans do when they are alone. Primatologists working in the field or with house-raised animals assure us that apes have a sense of their own role and status in society. Some of these accounts are fascinating—female chimps carefully arranging a mirror and then squatting over it while urinating, a chimp sorting pictures of animals and humans and placing her own portrait among the humans and well away from cats and dogs, but these are anecdotes.[12]

Consider the following classically simple experiment.[13] Chimpanzees, caged in isolation, were or were not permitted to see their own reflection in a mirror.[14] They were then anesthetized and their foreheads painted with an odorless, nonsticky red dye. People painted in the same way could not feel the paint itself after it had dried. The marks were located on the forehead where they could not be seen, except by looking in a mirror. After recovery from the anesthetic, the chimpanzees were again given access to their reflections in a mirror. The number of times they touched their face was counted. Those who were experienced with mirrors touched their red-dyed faces much more frequently than the others. The inference was that the chimpanzees who had learned their own appearance from the mirrors were concerned about having their appearance changed and were attempting to rub off the dye. Familiarity with mirrors had permitted the animals to develop a visual self-image which they seemed to defend against change. A small flurry of research attempted to determine how widespread among animals this property was. Self-referral behavior was found in all the great apes, with the possible exception of the gibbons.[15]

The full implications of this are not clear. It does not imply that great apes are stupid silent humans. It does, however, seem to imply that to some degree they have a kind of selfhood, involving some sort of sense of propriety, at least to the extent of rejecting the indignity of a painted face, if, and only if, they had previously realized that their face was not normally painted. They not only perform self-oriented actions, which might be considered merely analogous to a cat cleaning itself, but they can also take action to make reality conform to their mental image of themselves.

Their capacity to think about themselves is yet another filter between their environment and their responses to it. What they finally act on is sometimes, but not always, filtered not only through the sensory apparatus and through memory, which are found in all mammals, but through some kind of rudimentary thinking about their memories—an internal narrative. At any point in this process, simplifications are likely to occur, and conversely, there is even the possibility that sensory input will be complicated by becoming part of a mental scenario.

While the genetic distance between humans and chimpanzees is no greater than the difference between goats and sheep, the intel-lectual difference is very great. How great is this difference, and how and when it came about, is as fascinating a subject as can be imagined and cannot in justice be completely ignored if we claim to be concerned with how modern humans face intellectual problems. People have a "self," embedded in a social network and a pattern of social rituals that have been invented and accepted.

So far as I now know there is no clear evidence of ritual behavior among apes relating to the death of conspecifics. Chimpanzees show long-term symptoms of acute unhappiness at the death of their mothers even when they are old enough to have young of their own. It appears that the great apes certainly have strong emotions and may have a private self-image, but there is no clear evidence that they can share more abstract ideas, such as two animals mourn-ing over a dead friend, although the question is under investiga-tion.[16]

Investigation of the primate line that evolved into humans in-volves detailed analysis of fossil anatomy, paleoecology, geological dating, and the ongoing search for new fossil material. Despite the fascination of this research area, the only part of it that is vital to my immediate concerns is what little evidence there is on the mental activities of our more remote ancestors and their cousins. No neat set of texts exists to provide the history of the changes that occurred in the transition between the mental competence of our less clever early ancestors and our more clearly human recent ones. On the other hand, we have absolutely no reason to believe that mental capacity has materially increased during the last 20,000 or so years, which is the only period for which we have documentary evidence. Average human intellect, however one chooses to measure it, may have been constant for as long as the last 50,000 or even 150,000 years. Unfortunately, for the preliterate, preiconographic beginnings it is very hard to tell. Neither fossil skulls nor stony casts of brain cavities, which may retain their shapes for even millions of years, are records of thoughts and dreams, by which we recognize our intellectual humanity. Visions of the world, which is our chief concern, cannot be preserved but may only be inferred from artifacts or from careful study of continuing living traditions. Neither all traditions nor all artifacts survive forever. The language spoken to me by my grandparents has almost disappeared, along with almost everything of theirs that moths or mold could consume.

The East Bronx, where I grew up, is a ruin, its buildings burned and gutted and either fallen or waiting for the bulldozers, not a victim of war but rather destroyed by changes in social goals. During the mayoralty of John Lindsay, New York City's planners had a policy of "selective necrosis" by which regions of the city were permitted to decay to permit rebuilding. In this case the necrosis went too far.

I, therefore, feel I can imagine what the citizens of Roman Britain felt when the baths and villas had to be abandoned to decay, but because feelings are so ephemeral I know that my feelings are not knowledge. Documents and photographs and literature written by people with personal memories permit people without any personal contact to know a great deal about the city of my childhood,[17] but even if no verbal or pictorial record existed, examination of the ruins would permit an archaeologist to infer a great deal about the thoughts of those who lived there. Even now one can observe the ruins of special buildings for special functions—theaters, schools, libraries, synagogues, and churches—and special areas, like parks, with playgrounds, swimming pools, boating lakes, and fountains, and with walks lined with benches. An archaeologist would not know about the row of grandmothers sitting on the benches and keeping half an eye on the playing children.

While we can infer some aspects of mental processes from material remains, many material remains decay or are never found, so that we generally underrate the competence and intellect of their creators. We know from field observations[18] that chimpanzees strip extraneous leaves and projections off twigs and use the resulting tools to fish for ants or termites. But the twig tools, which are made quickly and discarded easily, although they may litter the ground where chimpanzees have been "anting," decompose within a few months. Without observing the living animals we would know nothing of their toolmaking.

Chimpanzees, in some places, also routinely smash things with rocks, and this may alter the rocks in recognizable ways, but in the absence of direct observation on the living chimpanzees it would be hard to tell from the appearance of the rocks alone whether they had been used by chimpanzees or by some other primate, perhaps even humans. Certainly hand-held rocks are still a valuable way of opening nuts all over the world and are indispensable for deshelling lobsters at the best of clam bakes. In fact, if we find rocks with flakes broken off of them in ways that are not likely to occur from being rolled by a flood or split by frost or from some other simple geological process, the assumption is usually made that the rocks or the flakes were once used as tools. In many places around the world the ground is littered with such stone tools. Stone, even more than metal, resists decay and preserves the scars of its own history, sometimes for millions of years.

Stone tools vary in complexity and in the skill and time required for their manufacture, but there is usually no question about whether or not a stone has been a tool—the two 500-year-old broken arrowheads I dug from my tomato patch were obviously arrowheads. Only the oldest stone tools, cobbles almost 2 million years old from which one or two flakes have been removed, must be thought about a bit before one can be convinced they really were tools. In some parts of the world the use and manufacture of stone tools continued into recent history. I saw in a museum in Aarhus, Denmark, an amazing flint dagger made by a stone worker less than 2,000 years ago in imitation of a metal dagger. The flaking had sculptured the stone so finely that it had the appearance of the seam that would be found on the sides of a cast bronze dagger.

Artifacts and fossils are dated by examining the material around them in or on the ground. This may consist of other artifacts—for example, bits of pottery of particular styles that had been dated from other locations—fossil animals or plants, and rocks, which can be dated by analyzing their atomic isotopic constitution. Once a rock is formed, the radioactive materials in it continue to decay, and by comparing the total amounts of radioactive materials and the amounts of the products of radioactive decay and knowing the rates of various radioactive-decay processes archaeologists can place actual years on the date of formation of the original rock itself. This has only been possible for the last few decades, but before that, while years could not be determined, relative dates could be found by noting which objects occurred in deeper deposits and which in shallower ones above them.

In Africa have been found fossil bones and artifacts of hominid primates, creatures which, if they were not ancestral to humans, were certainly more closely related to our ancestors than are any of the modern apes. The ages of these bones and artifacts range from 300,000 to 4 million years. These creatures, and the intellectual problems and disputes that they are engendering are described in several fascinating technical and popular books.[19] These African tools and fossils provide a glimpse of millions of years of history, permitting some tentative insight into the question of when, and to some degree how, the human capacity to complicate and simplify appeared.

Stone tools are not randomly scattered over the landscape. They are clustered in small areas on the surface or in caves that may have been camps, or sites where large animals had been slain and butchered, or perhaps places where floods have left them. Each cluster of stones usually contains several types of apparent tools and also small flakes and bits of broken stones that don't seem very useful. The crudest tools are generally found in the oldest collections. While they clearly could not have been produced by purely geological processes, many of them could have been the result of simply picking up a handy rock and banging it on another rock several times, in the way of modern chimpanzees. But even in the oldest assemblages of tools there are some whose flaking pattern can be explained only as being the result of repeated striking of the stone in one relatively small area. In every assemblage some tools also show a definite, sharp working edge that could be produced by a repeated series of blows, not necessarily in any particular order but all localized at the edge. These might involve blows all from one direction (unifacial) or from both sides of the blade edge (bifacial).

The overall shape of the tools in the earliest examples is essentially unmodified from the original rock, although they have either a working edge or a burin or awl-like point. In later specimens the tools have been repeatedly chipped, producing characteristically shaped classes of objects. Hand axes, shaped like flattened pears, fit conveniently into the fist and are apparently easier and more lasting when used to butcher a goat than are the less carefully shaped but equally sharp-edged stone fragments.[20] Others have had flakes removed so that what is left behind is an almost perfectly round, more or less flattened discoid; others have a rectangular shape and a chisel-like sharp flat edge.

Most stone tools seem to have been made rather quickly—a matter of minutes—and used until they lost their edge. Then either they were sharpened by striking off the old edge for a new one or they were simply discarded. Some tools required that only a half dozen flakes be removed from a cobble, while some of the later ones were so completely reshaped that the original stone shape is essentially obscured.[21]

Various attempts have been made to infer the intellectual capacities of the makers from the nature of their stone tools.[22] Wynn, for example, has asked: What are the simplest spatial concepts that would have been necessary to manufacture the different types of stone tools found in East Africa?[23] How did their concepts of space compare with our own, in which we can visualize objects in three-dimensional space and even visualize how they would appear from some viewpoint other than our own? Adult apes and sufficiently young human children do not seem capable of this degree of mental freedom. He concludes that the tools from 1.5 million years ago give no evidence that their makers focused on anything other than the tool edge. In the tools of approximately 1.3 million years ago he sees indications of concern for the overall shape of the tools as such, in addition to concern for the edge. In tools from only 300,000 years ago he infers that the makers must have visualized the shape and symmetry patterns of the final tool while working, even though that shape was not readily discernible in the stone itself. He concludes that the tools of more than 2 million years ago showed no more profound a sense of space than the tools of modern apes, while "by 300,000 years ago hominid reasoning ability had become essentially modern."[24]

But I am really more interested in what these hominids did with their reasoning ability. When did they begin playing games with some intellectual content? Did they wait until they could match wits with a modern engineer, or did they begin before they had such intellectual power? Young children save pretty things, dress up, undress, sing songs or keep silent, before they attain human reasonableness. Of course the greater the intellectual power the richer the potential world of make-believe, as we see all around us, but when did the world of imagination begin?

Paleolithic and neolithic art stands as equal in talent and imagination with any modern art, but this is merely tens of thousands of years old. Burial rituals are maybe four or five times older than that, and the use of fire twice as old again. But there is evidence that around 3 million years ago a reddish cobble, which had been eroded so that it seemed to have figures of faces on it, was carried about as "a treasured object" by a South African hominid. In an Ethiopian site occupied by the more-nearly-human *Homo erectus* 1.5 million years ago were found pieces of weathered basalt which produced red pigment when rubbed. A treasure of red ochre from 300,000 years ago was found at Terra Amata near Nice. Oakley suggests that "interest in mineral-red pigment [may mark] an important threshold in the hierarchy of evolving human minds. Did it perhaps coincide with the beginning of social organization? . . . Man may well have turned to artificial coloring of his skin and apparel as soon as tribes, societies, and family groups were formed and required to be distinguished when sighted at a distance" (p. 207).[25]

This certainly seems plausible, but I prefer to think of colored images and painted faces as props in a story about the world. The consistency and elaboration of stories may well be necessary before one can believe in tribes and families or feel at home in the world, as we will consider in the next chapter.

The capacity to speak, which evolved fairly late, increases the possibility for self-awareness by an order of magnitude.[26] Not only do people have a normative introspective self-image by the time they are three years old, but they can talk about it and create for themselves a matrix into which further information, narrative, opinion, and mental constructs are incorporated as their experiences accumulate. Although an examination of the origin of language in the ancestry of humans would make this discussion entirely too long and even more speculative, it seems self-evident that the development of language, particularly nouns, would co-occur with the beginnings of storytelling because nouns permit a story to include both the immediately observable and things that were either not present or nonexistent. Nouns enlarge the richness of the self-image of both speakers and their audience. They permit internal images to be compared.

I see no way of discovering the first narrative about the world.[27] Clearly, however, ornamenting a human grave, or even laying out a deceased pet hamster in a cotton-wool-lined cigar box, is an acting out of a story that extends beyond the empirical. The story does not imply a consistent or well-developed religious system. It does, however, imply some sense that objects have meanings beyond their obvious utility. If it were necessary, for some curious legal reason, to draw a clear line between human and nonhuman—for example, if a group of australopithecines were to appear and one had to decide if they were to be protected by Fair Employment Laws or by the ASPCA—I would welcome them as humans if I knew that they were seriously concerned about how to bury their dead.

Research workers intensely argue about the tall, strong, large-brained, heavy-jawed people called Neanderthals. Were they ances-tors of *Homo sapiens,* or were they a divergent evolutionary lineage that shared a common ancestor with *Homo sapiens?* More interesting for my present purpose is a recently discovered grave made by Neanderthals more than 50,000 years ago in Turkey. The soil around the grave is laden with flower pollen, as if bouquets had surrounded the corpse.[28] So far no earlier evidence of burial rites in prehuman hominids is known, but several more recent Neanderthal graves exist, and with the appearance of *Homo sapiens* around 30,000 years ago ritual burial sites became reasonably abundant.

I believe that ritual burial demonstrates a fairly elaborate con-structive theory of reality. Further, I believe that statement to be objective and free of my own personal bias and culture-bonding. Sometime earlier than around 50,000 years ago at least one species could exchange ideas. Without going into whether or not *Homo sapiens* are descendants of Neanderthals, I can at least say that sharing ideas is one of the most important qualities of humans and that we share this property with Neanderthals.

I believe that the evolutionary process has provided us with the capacity to share ideas without imposing constraints on our ideas. Preformed decisions about how to deal with complexity *correctly* are not part of human nature. In fact, human freedom from evolu-tionary constraints combined with the persistent necessity of making decisions lends interest and danger to political and social life. Think how much easier it would all be if we had "instincts" to rely on! As it is, we must rely on our "self," which is by no means a simple assertion.

The nature of self is a recurrent intellectual theme. In a recent philosophical analysis Taylor defines self in terms of the necessary framework within which each human functions.[29] He says, "Doing without frameworks is utterly impossible for us . . . living within . . . strongly qualified horizons is constitutive of human agency . . . stepping outside these limits would be tantamount to stepping out side . . . undamaged human personhood" (p. 27) . . . "Our identities, as defined by whatever gives us our fundamental orientation, are in fact complex and multitiered . . . the human agent exists in a space of questions . . . to which our framework definitions are answers providing the horizon within which we know where we stand and what meaning things have for us" (pp. 28–29).

While I find myself in wholehearted agreement with these assertions, I was surprised by Taylor's deliberate rejection of evolutionary and comparative information and his insistence on the absolute, rather than quantitative, uniqueness of humans. Taylor refers to the chimpanzee mirror experiments briefly but rejects them as being "obviously" unrelated to his concerns with the human self. I find this by no means obvious. To discuss, as Taylor does, a supposedly uniquely human property—a sense of selfhood—which is innate or necessary (that is, biological) without at least considering information from a sibling species that shares 98 percent of its genes with humans seems curious. To find no trace of this property in such a close evolutionary relative would be surprising to a biologist and would strengthen Taylor's position about the uniqueness of humanity.

Perhaps Taylor's position merely indicates his distaste for natural scientists' intrusion into what he sees as a philosophical problem. In fact, his distaste is not entirely unjustified. No sound evidence has been found to support the hypothesis that politically significant virtues or vices are attributable to any sort of hereditary human nature. This question was at the heart of the "sociobiology" battles initiated by E. O. Wilson and Richard Alexander during the 1970s.[30]

The genetic arguments were based on what particular genes might be expected to do if they existed, rather than on any actual demonstrated extant genes. After a lot of vituperation on several sides, what became clear from this controversy was that, at very best, it did not seem possible to demonstrate for any primate—and certainly not for humans—any relevant genetic constraints on ethical or, more broadly, political decisions. Specifically, it has not been unequivocally demonstrated that a particular human individual will favor those individuals with whom a greater number of identical genes are shared. The failure to demonstrate a genetic-similarity-driven affection on the personal level, or patriotic sentiment on the public level, or conversely a gene-difference-driven personal revulsion or xenophobia in no way denies the common observation that common culture is a major force in permitting amicable behavior. Cultural differences, including differences in language, dietary preference, social and sexual customs, along with economic and educational differences, seem plausible explanations for various forms of bigotry, xenophobia, and group antagonism, without invoking genes at all.

Often correlations exist between physical appearances that are indeed traceable to genetic differences on one hand and cultural divisions on the other. These obscure the possible significance of genetic differences having any relation to sympathies or antipathies. In short, the presumably counterrevolutionary behaviors of altruism and kindness, which motivated much of early sociobiological speculation by Alexander and Wilson and their students, have not been demonstrated to follow genetic constraints. Altruism and kindness transcend genetic relationships, as Androcles pointed out to the lion.

My arguments for rejecting the sociobiological model of genetic control of human behavior, and for being extremely doubtful that appropriate sorts of genes will ever be discovered in humans, have been published elsewhere.[31] They have nothing whatsoever to do with political considerations. Rather, they are based on my conviction that the intervening processes between the genotype and decisionmaking in humans are uniquely complex and make it extremely unlikely that ethical or moral behavior has anything to do with measurable genetic differences—except in the special cases in which metabolic disorders contribute to what is considered antisocial be-havior, as in porphyria or possibly alcoholism or conceivably some of the mental disorders that are called schizophrenia. Even here the definition of "antisocial" is culture-bound, as can be seen from the existence of well-defined roles in some societies for persons that would be considered antisocial or even evil in others.

Another reason to distrust the sociobiological model is because, like Freudian analysis, it can explain entirely too much too easily. No scenario is inexplicable. For example, in a sociobiological reading of Shakespeare's Hamlet, Claudius' unsympathetic attitude toward Hamlet may be seen as blatantly equivalent to the killing of preex-isting young rats by a newly arrived male. Claudius obviously has less genetic investment in Hamlet than in any progeny that might arise from his relations with the queen. Queen Gertrude's ambiva-lence toward Hamlet is reminiscent of the process of abortion in female rats that have found new mates. Laertes' surprisingly mild response to the death of his father Polonius but very dramatic response to that of his sister Ophelia is easily explained by the low age-specific reproductive value of Polonius compared with the very high reproductive value of Ophelia. I leave as an easy exercise the process of explaining Hamlet's obsession with his mother's sex life—this can be done optionally in Freudian or sociobiological format—but neatness counts.

While human life might have been simpler to explain had the sociobiological hypotheses been valid, no genes have been found, which means that at least for now we are left with the observation that humans—and to a lesser degree modern great apes—behave in the context of an intellectual construct in which they place them-selves in what Taylor calls a "field of values." What those values are for any individual, how aware individuals are or are not of the existence of the field, and whether there are any restrictions on the values in the field seem like open questions from the standpoint of scientific analysis, although in the human context they are the substance of moral philosophy as well as practical morality.

I am agreeing as strongly as I can with Taylor's insistence on the existence and importance and inescapability of individuals' living in a field of values. Information that makes its way through all the filters may be internalized into this field, where it must somehow

enter an internal narrative in some way, even if this means rear-
ranging the field itself. Human actions, private and public, are
performed in the context of the values of the individual. What
those values are varies among cultures and individuals. What they
ought to be is a problem for moralists. What will concern us are
creations within this field of values. That is why we had to consider
the origins of the self. A sense of self is a prerequisite to the human
intellectual capacity to deliberately simplify or complicate.

2

Simplifying Religious Revolutions: The Birth of Doctrine

Self-image, a prerequisite for those intellectual constructs that can be deliberately simplified or complicated, is intimately intertwined with deep and sometimes very private beliefs about the nature of the world. When beliefs are communicated, the result may be a shared, common set of beliefs, and these in turn may be institution-alized eventually into religious doctrine. In this chapter I will con-sider the doctrines of organized religion. This aspect of religion is conceptually distinct from deep beliefs on one hand and from life-styles on the other (see Chapter 9). Deep beliefs need not be ex-pressed as doctrines, and life can be lived without conscious doc-trinal justification. However, in the process of articulating deep beliefs into religious doctrines, matters of complexity and simplicity have an important role, as can be seen by brief historical review of both Eastern and Western religions.

At some point, lost in prehistory, stories began. I believe that some of these stories were ancestral to doctrines. Some stories were purely for entertainment, while others purported to describe reality. In this latter group were some that dealt with obviously practical tasks, and these had to be internally consistent if the tasks were to be accomplished. But many tasks involved processes that were not readily apparent, and the stories attached to the tasks described this part of these processes also. How to make a fishnet and how to set it and draw it in required communication, if only by example. But isn't there more to catching fish than that? Why did the fish come to the net? How can we say thank you for the fish that have come, and how do we complain about their absence, if they are absent? We know that we place seeds in the ground and eventually gain a harvest, and how and when to do that is important knowledge about the world. But what happened to the seeds? How did the seeds grow? The effectiveness and veracity of explicit communica' tion about how to plant seeds and set nets was directly testable. But stories about the unseen aspects of reality could be invented freely without interfering with practical affairs.

Nevertheless, stories about the unseen and untestable parts of the world may be more serious than stories told just for fun, which can be changed to suit moods or to make the audience laugh or cry. Up to a point they are honest attempts to build a consistent theory of the real world. They are not counterfactual in the sense of denying known empirical information but rather nonfactual in the sense of not having a close connection to empirical tests. By the time we find what we now speak of as "religions," imaginings about reality had far outstripped in richness and complexity the available empirical information.

From that point on, exploration of the empirically accessible world seems to have separated from exploration of what seemed to be the deeper reality beyond the empirical. Theories of deep reality took on a life of their own and in various ways exceeded the standards for intellectual and behavioral simplicity. Around 3,000 years ago in the literate world attempts were begun to maintain the essentials of an understanding of reality by minimalizing the older stories. Each time this was done, new and different elaborations grew out of the minimalist base and engendered new minimalistic "corrections." For thousands of years there was no reason for this process to end.

The first use of writing was to preserve correctly the marvelous stories about the unseen aspects of the world, along with business documents and the boastings of kings. But writing is too permanent to contain the changes from year to year and generation to genera' tion that occur in living imaginations. There is always the possibility of deviation between the written, officially recognized account of the hidden world and that which is responding to changed circum' stances. Around every authorized account is a buzzing swarm of unauthorized accounts. Today, for certain classes of people, the authorized account is the story as printed in the *New York Times*. I have found that whenever I have first'hand knowledge of an event described in a *New York Times* news story, I find mistakes in the newspaper. Despite this, I tend to believe almost uncritically the stories about which I have no personal knowledge.

Sometimes assertions about the hidden nature of reality become so formal and so widely accepted that they reach the status of religious doctrine or dogma.[1] Around any such doctrine is a cloud of variants, misunderstandings, and crumbs of outmoded earlier doctrines which, until they deviate too far, are harmless and even charming. But most systems of religious doctrine, regardless of their other differences, share the assertion of their own validity, so that if variant opinions become so coherent that they constitute an alternative doctrine, then the two are on a collision course. In short, the process of changing doctrines occurs over the objections of the supporters of the established doctrines.

The flexibility of belief systems has a limit. Advocates of suffi-ciently great doctrinal changes are rebellious, and serious attempts to deny existing authority are rebellions. The ones that succeed to the point that they themselves must watch out for new rebellions are called revolutions. Since important political and conceptual rev-olutions, as opposed to minor coups, involve transitions in deep belief systems, the history of revolutions may be the most difficult kind of history. Even the participants may be unaware of what they are part of.[2]

Rebellion usually carries risks and therefore is not undertaken unless an intolerable distance between deep beliefs and established doctrines has developed. What is, or is not, intolerable and whether or not the intolerable leads to rebellion depends on circumstances. I once stood with a friend in a long queue for tickets to a second-rate film. He became loudly annoyed by the wait and then stopped himself, saying how utterly unimaginable it would have once been to worry about a film queue. My friend was a Dachau survivor. Primo Levi reports that there were essentially no rebellions, and certainly no revolutions, in the German concentration camps during the Second World War. Where there is no hope, there is no rebel-lion. The best-fed prisoners led the few rebellions that did occur. Similarly, the slave rebellions of America were often led by the most privileged slaves, and apparently the French, British, Russian, and American revolutions were all initiated by the relatively prosperous, rather than the completely downtrodden.[3] If only the physically

well-nourished can undertake rebellion against the physically intolerable, one could argue that only the intellectually well-nourished
can undertake intellectual rebellion, which may account for the
rarity of successful major doctrinal revolutions.

To attempt to change the world, the rebel must have some vision
of some other world. Doctrine, lifestyle, and the field of values that
define the self are so intertwined that all serious intellectual revolutions do have the effect of changing, or even creating, deep belief
systems (see Chapter 1). Conversely, counterrevolutions can cause
reversion to patterns that resemble those of prerevolutionary time.

The French Revolution attempted to substitute what was
thought to be natural rationality for medieval superstition. It introduced a new calendar, purged of the archaic month names, and the
"natural" metric measurement system, in which a meter was one
millionth of the distance from pole to equator. During the 1920s
the Soviets briefly experimented with a ten-day week, and corresponding alterations in worldview were desired by the proponents
of the Chinese Great Leap Forward and by the recent Islamic
revolution of Iran. None of these revolutions was permanent. The
recent dismantling of Communism in Eastern Europe seems to be
accompanied by attitudes and behaviors reminiscent of those of the
prerevolutionary nineteenth century. These examples only indicate
that revolutionary doctrines, or perhaps doctrines in general, need
not be tightly connected to deep belief systems or ideas of self.

Some periods show more significant doctrinal revolutions than
others. The first millennium before Christ was a time of revolutions
in Greek, Jewish, Indian, Mesopotamian, and Chinese doctrines. A
world conference could have been held in 550 B.C. at which the
molders of much of the thought of the next 2,000 years could have
met, quarreled, and put out a symposium volume. This curious
temporal juxtaposition was the basis of a fascinating modern symposium and a best-selling novel.[4] Had such a meeting actually occurred in the ancient world, and had a single consensus been reached
about the nature of reality, think how much intellectual richness
would have been lost. As it happened, different schools of thought
had developed institutionally vested interests long before disciples
of these views met and had to face the problem of arriving at
consensus. The resultant quarrels continued to be important until

modern science swept the field, and in some ways the disputes have not yet been settled.

Why was the first pre-Christian millennium so fruitful in doctrinal revolutions? One suggestion is that as different peoples were thrown together in trade and war, an uncomfortable inconsistency in mythologies was recognized. This may not have bothered many people, but Herodotus, arguably the first historian, seems very concerned with the fact that different peoples have apparently different pantheons, and he was at pains to reconcile them. Which Egyptian god corresponds to the Greek Zeus, or Athena? Since not all the gods of the different pantheons seemed to translate conveniently, several gods from one system could be considered to be different aspects of a smaller number of gods in some other system—thereby developing syncretic deities that "included" others.

One might even consider a single god that contains an entire pantheon—which Nikipowetzky refers to as a "henotheistic monotheism." He strongly distinguishes this belief from monotheism, which requires a deep belief that the unity of god is such that it is simply pointless to consider seriously any arguments at all that discuss how many gods exist. Ethical monotheism in this sense is considered to be a peculiarly Jewish contribution, arising out of the political split between Israel and Judah, which demonstrated the independence of religious belief from specific governments, and out of the Babylonian exile, which demonstrated the independence of religious belief from geography. Stating these prerequisites in no way diminishes the distance of the intellectual leap to ethical monotheism itself.[5] In any case, the great simplification of rejecting the entire pantheon, or its relegation to the rank of a mere story rather than the basis of deep belief, was at the heart of the message of Buddha, Mahariva, and the great Greek philosophers, as well as that of the Jewish prophets.

We know of these early thinkers from written texts. In most cases texts that are the bases of successful intellectual revolutions seem to become encased in orthodoxies, just as innovative artistic approaches harden into academies. It is as if original thought is extremely tender and painful to contemplate until it is encased by a doctrinal shell.[6] Throughout literate history attempts at maintaining revolutionary fervor have crystalized into catechisms and creeds. Approximately two generations are required for this process, necessitated perhaps by the death or weakness of the founders and their immediate disciples. *The Communist Manifesto* and *Das Kapital,* Mao's *Little Red Book,* the *Book of Mormon,* and the *Green Book* of Kadhafi are modern examples. The Old and New Testaments of the Bible, the Koran, Upanishads, and Vedas are older examples. In every case they appeared after the ideas of the revolution had already been clarified to the leaders' satisfaction and were in danger of being corrupted.

These are obviously serious and important texts. One cannot take liberties with them as one does with casual anecdotes or purely literary creations that are seen as entertainment. Also the fact that they are either written down or embedded in a firmly crystalline oral system of transmission—as in the genealogies of the Maori, historical songs of the Ashantis, and various kinds of family oral histories—prevents their being treated casually and inhibits inno-vation. Often the written texts represent the termination of an oral transmission system, as in the Vedas and parts of the Bible. In any case, this kind of story has a special character. It is "sacred." Students may be required to memorize, interpret, or comment on the Bible or Koran or *Little Red Book,* but they are not asked to correct or improve it or to imitate it in an original composition of their own.

Each major religion represents a revolutionary attempt to grasp some kind of new reality, usually embodied in a story of a single nominal leader around whom coalesced a body of followers. The coalescence process typically involves a rejection of some preexisting doctrine. The difference from the accepted system had to be em-phasized if the new system was to be more than a minor variation within the old scheme, even if the leader had no such intention.

A new religion involves radically modifying an established belief system, whether or not the founder is conscious of his revolutionary role. While founders of religions are often seen in retrospect to be revolutionaries, in fact we are usually ignorant of their true initial intentions. Martin Luther revolutionized Western Christianity, but he thought of himself initially as a reformer within the Catholic Church. Henry VIII of England in fact founded a Protestant Church, but he held his title of Defender of the (Catholic) Faith throughout his life. The accounts of the life of Gautama Buddha show him as a reformer within Hinduism.

In the establishment of the Mormon Church, the founder from

the beginning of his ministry seemed aware that he was starting a revolution, but in the world of Joseph Smith religious revolution was almost the norm. Mormonism was one of many sects founded in the "Burned Over District" of western New York state.[7] The district received its name from the fires of enthusiastic missionary activity that it had been exposed to for the first four decades of the nineteenth century. Religious rebellions were as common in that region as new dietary fads were in California in the 1960s. Most of the rebellions died out or abandoned their often very curious beliefs and practices. We will return to these later, in the context of simplicity as a way of life rather than as an intellectual game (see Chapter 9).

In literate societies the object of the revolt is to replace a body of beliefs that has lost most of its own revolutionary character somewhere in the past and has been hardened into a creed, set of practices, series of texts, and fixed attitudes. This establishment constitutes a sitting target. A revolutionary idea, on the other hand, is a kind of intellectual guerrilla force and therefore almost immune to direct attack, until it too crystallizes.

In most cases our present sense of the history of a great religious movement is based on its own sacred texts. The need for these texts arises after the revolutionary phase has passed, often centuries after the life of the founders. This is true of Judaism, in which the Pentateuch was put into canonical form in the time of Nehemiah and Ezra, in the sixth century B.C., a thousand years after the time assigned to Abraham. The New Testament narratives of the life and times of Jesus appeared at least a century after he died. The same is true for Buddhism, Jainism, Confucianism, Taoism, and Hinduism. Even the Koran took almost a century after Mohammed to become a canonical text. Only very recent and relatively minor Protestant sects, within the general world of Western Christianity, have had the benefit of contemporary accounts at the time of their founding.

An actual belief system differs from a dogmatic creed and the ritual that surrounds it in that the words of a creed and the way it is copied or chanted have been canonized, frozen into a correct shape. Prior to the canonization process, verbal or pictorial statements of reality differ, with each version having some virtues and some faults, without any clear way of deciding which is correct, or even any clear sense of what "correct" means. One reason canonical texts are needed is to deny validity to other texts by declaring them noncanonical. The Nag Hammadi codices, preserved in a clay jar in Upper Egypt from the second century A.D. until 1941, include "Gospels" of James and Thomas and Philip, apparently written by self-confessed Christians but rejected by the branches of primitive Christianity ancestral to Western Christianity.[8]

Even if there are no rival texts, canonizers may be attempting to purify the body of elaborations that tend to encrust the remembered stories about the founders. The imagination of the storyteller, the desire to maintain audience interest, and the need to maintain a clear sense of the virtues of the founder against the vices and errors of his opponents require that the original stories be added to and each part embellished. These embellishments certainly produce something other than scholarly history and will produce variant texts and may even generate "heresies." The canonization of religions in sacred texts discards wholesale segments of the embellished creeds and attempts to restore a minimalist essence, but this is undermined by their followers who, in their turn, use the canonized texts as a base on which to elaborate.

Religions of the Book

This can be readily seen in the relation between Judaism, Christianity, and Islam. These three constitute the Biblical religions, what Islamic teachers call the "Religions of the Book." Each of these minimalized some aspect of preexisting belief, then elaborated some other aspect and in turn became vulnerable to new minimalizations of its most elaborate parts. Judaism began as a revolutionary minimalization of Greco-Persian theology, developing in its turn an elaborate legalism. Christianity minimalized Jewish legalism and permitted theology once again to become elaborate, and Islam minimalized Christian theology, developing a legalistic system of its own.

Before I briefly outline how this occurred, I should acknowledge that my relation to the different religions varies. I am at home in Judaism. However strongly I may disagree with particular aspects of current Jewish politics, social structures, or even doctrinal positions taken by "leaders," my role is still that of a participant. I am comfortable with Jewish liturgy, can participate in religious study groups, and can struggle through the appropriate literature without the need for a translator. I feel free to criticize Judaism as I criticize relatives or certain portions of science.

As an American graduate of a Campbellite college I have a reasonable familiarity with several sorts of Christianity but not the insight of a participant. For the other religions that I will mention— Islam, Buddhism, Hinduism, Jainism—I have only read translations and comments by other outsiders. By contrasting my own "inside knowledge" of Judaism with what I would have known if I had read only translations and outsider's comments, I know how shallow my own understanding of other religious systems must be. There-fore, despite the fascination of the subject matter, I will not make any attempt at a complete presentation of the doctrine of any reli-gion but will focus only on what I understand of the role of mini-malism and elaboration in each of them. Perhaps by staying within these narrow bounds I will avoid increasing the world's supply of misunderstandings generated by scholars describing religious systems from the outside and by apologetic partisans of particular religions writing from the inside.

The canonical texts of the Torah, the five books printed at the beginning of modern editions of the Old Testament, were prepared in a world dominated by polytheism, as manifested in Egypt, Greece, and Persia. There was an entire population of named deities, each with its own assigned function. Also, for most of its history the Persian Empire took a tolerant and syncretistic position with ref-erence to local gods. The various named gods were seen as differing manifestations of the same basic idea, and each city or state was entitled to its own. Against this elaborate system Judaism set a God that was the ultimate single nonsyncretic deity, so rich in attributes that he could neither be named nor envisioned. All the other gods were simply misconceptions of benighted heathen.

Even now, after 2,600 years, theology is not an interesting subject to most learned and orthodox Jews.[9] At the center of Judaism's wonderful richness of legalistic elaboration, legends, and poetic spec-ulations about unseen forces, there is a curious silence about the nature of God. I have heard one modern Protestant theologian innocently reiterate the medieval complaint that Jewish informants must be deliberately hiding their deep beliefs from him. They were all extremely reluctant to discuss the nature of God, and the little bits that were said or hinted at seemed mutually contradictory. What he could not accept is the Jewish concept that the entire subject is entirely too difficult to deal with, that reality in an ultimate sense really is inaccessible, and that therefore attention must focus on human behavior.

Religious "activity" within Judaism involves the elucidation of textual material so as to permit conclusions about real and hypothetical problems of a very practical nature. Not only is there essentially no focused concern about the nature of God, except by exclusion, but there is very little explicit concern with the nature of such abstractions as truth, justice, or sin. Defining justice in the abstract is considered much less significant, in fact a waste of time and effort, compared with discovering how justice should be applied in particular cases, how liturgy should be conducted, and how to behave properly. Study of texts that relate to these problems is what traditional Judaism means by "learning."

The act of learning focuses, directly or indirectly, on the set of volumes called the Talmud—an amazing elaboration of intellectual religiosity. The Talmud cannot even be printed in a simple way. On every few pages are a few lines taken from the Mishnah, a sixvolume treatise describing the operation of the legal and liturgical system during the time of the Jerusalem temple, completed in the second century. Following the bit of Mishnah might be a few paragraphs or a few pages of Gemorrah—a kind of *precis verbaux* of the rabbinical academies of Babylonia or Palestine, in which the Mishnaic text is explained and clarified to a fare-thee-well. No word is left unprobed if there is in anyone's imagination a way that it might be misunderstood or if there is any association that anyone might have with the text that might be of any intellectual interest. Around the margins of the page, one outside the other, are columns of commentary, some of it very recent, explicating the explications. Printed in a form that seems to protest the two-dimensionality of the printed page, it is a book arranged like the layers of an onion or the leaves of a cabbage. It may be consumed in slices, or leaf by leaf, and one may start from the middle, the Mishnah, and work out, or with the latest commentators and work in.

In one amazing section of the Babylonian Talmud, the discussion starts from the Mishnah on whether or not it is permitted to repair a fence on the intermediate days of an eight-day festival. The Gemorrah quotes a pre-Talmudic text, now lost, establishing that a fence to keep out rats and *aishet* can be repaired but not begun, but *aishet* is a rare word which needs explaining and a text is dutifully uncovered defining *aishet* as the Palestinian mole rat, among whose properties are the tendency to hide from the sun and to dig into, and disturb, ants' nests. This reminds one rabbi of a "great way" to disturb ants nests—by mixing ants from different nests or adding the dust from one nest to that of another. The commentator Rashi, around 900 years later, asks why "dust" rather than simply dirt. Because he says, on another authority, the ants fight because of the odor of the nests being different and the odor is obscured if the earth is wet! (Pheromones—odoriferous chemicals that regulate animal behavior—were "discovered" in ants in the 1960s, and we may expect some future edition of the Talmud to add a discussion of pheromones to the outermost columns of the appropriate page.)

I can't resist one more example. What is to be done with a book (a scroll) that one has found? Obviously, justice requires that it be returned if the owner is known. What is more interesting is how one should relate to the book while waiting for the owner to show up. Is one permitted to read the book, even though it is not the finder's property? If a book is not read it might grow moldy, so that occasional reading is good for the book itself and therefore to be encouraged. But what if it is a difficult book, of the sort that should not be read alone under ordinary circumstances lest one develop erroneous ideas which a companion might correct? Can two people read it together? There is the danger that they may read at different speeds and both become so engrossed in the page that the book might become torn as they pull on it. Therefore reading together is not advisable. And so on for many dense volumes.

The minimalistic theology at the center of Judaism has permitted the development of legalistic and literary elaboration and the creation of a great intellectual game. Even now not only do devotees give most of their waking hours to the Talmud but also study groups spring up all over the world whose members take pleasure in spending an entire evening gaining some understanding of two or three paragraphs of archaic Aramaic text, playing at being scholars, and then return to their jobs as salesmen, engineers, businessmen, or what have you.

Where theology might have been in Judaism is a melange of poetic images, metaphors built on other metaphors, impossibilities and extravagances intended to boggle the mind. This is the Kabbalah, a poetic, and sometimes theological, theosophical, theurgic, and ecstatic obverse to the Talmud's literalism and intellectuality. Here one finds the full panoply of powers, angels, and heavenly forces but usually understood with a disclaimer to the general effect of "We are discussing that which cannot be discussed. Heaven forfend that any of it be taken literally, but the discussion is nevertheless of monumental importance."[10]

Questions about the nature of God are essentially forbidden in the legalistic mainstream of Judaism, although the Kabbalah finds similar questions suitable starting points for poetic speculation. In the Kabbalah of Isaac Luria of Safed in the sixteenth century the notion of the world having been created by divine power cracking the ten concentric spheres surrounding the Ein Sof, the Endless, provides a kind of answer to an unasked question about the nature of God.[11] It also gives a curiously neo-Platonic explanation for good and evil by producing a world in which sparks of divinity are encased in shells of baser stuff. The obvious questions about immanence and transcendence, about duality in creation, and so on which could have fueled endless theological congresses were not asked, nor was anyone excommunicated. This was not only because it was the sixteenth century, which was too late for that sort of thing: in the first and second century equally fantastic questions and answers were being supplied in proto-Kabbalistic works. Rather, Kabbalah is taken as a kind of very serious poetic game that actually regulates behavior to some degree in the sense of providing a script for right intentions, a path to intense personal pleasure, and occasionally a guide to action. Some of the same people that are masters of Talmud and law also study and enjoy Kabbalah. My grandson's other grandfather, an absolutely charming man who has been a French civil magistrate, head of a rabbinical court, and municipal rabbi of one of the major cities of Israel, requested that my grandson's name be chosen in accord with Kabbalistic suggestions.

Christianity was founded a half millennium after the writing of the basic biblical texts, which were already surrounded by the elaboration of rules, ceremonies, and restrictions, out of which would appear the Talmud. Christianity at its founding did not immediately focus on theology. It did advance a new minimalistic transformation of Judaism. It seems apparent from the Gospels, quite apart from scholarly arguments and concordances, that Jesus denied the legalistic spirit of Judaism. By eating what was forbidden, by associating with pariahs, by avoiding political protest, and by healing a chronic medical condition on Saturday, Jesus focused on the *ideas* of justice, charity, piety, and so on rather than on the minutae of their application. Like most minimalistic revolutions, Christianity swept away massive complications in a preexisting system and focused on an essential idea that the preexisting system took for granted.[12]

Exactly what happened next and why is not really clear, but within a few centuries of the writing down of the Gospels, Christianity had developed its own system of elaborations, not legalistic but theological. That is, the New Testament, in part written when most Christians were former Jews, treats theological problems with Jewish lightness. Later, when most of the members of the Christian Church, instead of being onetime Jews, were onetime "Greeks" (that is, Hellenized Middle Easterners), explicit and serious questions were raised about the nature of God. If the godhead was a trinity of Father, Son, and Holy Ghost and if the Son was Jesus who was born to a woman and died, was there a Trinity prior to the birth, or perhaps the death, of Jesus? If so how could Jesus be a man in any biological sense, and if he was not a man what was the meaning of the crucifixion? If Jesus was a man while on Earth, how could it be asserted that God was immutable and unchanging, since the nature of the godhead must have switched from a unity, or perhaps a duality, to a Trinity, at a historical moment?

The Nicene Council of the fourth century, after long debate and deep political infighting, came out with a very carefully worded compromise statement. The Son was "of the same substance as the Father, begotten not made . . ." This is not merely archaic language but rather a very careful attempt to articulate a deep truth about the nature of reality. Despite the fact that the Nicene Creed was nominally accepted by all but the Arians and to this day is printed in many Catholic, Orthodox, and Protestant Christian prayer books, the problems of the nature of God and Christ had not been settled.

In fact, the Nicene Creed was one event in a history of bitter theological dispute that extended from Marcion of Sinope in the early second century, who tried to explain the existence of evil by distinguishing between a benign Pauline God with no responsibility for evil and an old creator God who lacked moral sense, to Severus of Antioch 400 years later, who was excommunicated for denying the limited sense of the duality of the nature of Christ which he perceived in the conclusion of the Council of Chalcedon. That council, in 451 A.D., described the nature of Christ as follows: "Born of the virgin Theotokos as to the manhood, one and the same Christ. Son and Lord, only-begotten, made known to us in two Natures, unconfusedly, unchangeably, indivisibly, inseparably. The difference of the Nature being in no way removed because of the Union, but rather the properties of each nature being preserved and concurring into one Prosopon and one Hypostasis." Frend notes that this is a profoundly technical wording, comprehensible only in the context of the theological theory of the time.[13]

The seriousness of these theological questions is made evident by the anathemas and counter anathemas, excommunications, and schisms that have characterized Christianity to this day. The fourth-century Donatist advocacy of the separation of church and state is still an issue in the United States, as is the Montanist assertion, of the second century, that there is validity to contemporary prophecy in the spirit. The assertions of Cyprian of Carthage in the third century that priests must be holy in their private lives has recently echoed in a media circus in the context of television preachers. Who was permitted to redefine orthodoxy and who was spun off as a heretic, which churches lived and which vanished, which problems remained interesting and which were simply forgotten were obviously results of complex historical and personal factors as much as they were the results of purely technical theological disputes. What is clear is that Christianity in its first half millennium took theology very seriously, as if theological errors really damaged the soul of individuals and polluted the world. The notion of deity had not been complicated into a pantheon, but within a kind of monotheism a most elaborate intellectual structure had been created.

The region bounded by Carthage, Rome, and Constantinople, the western border of Persia, and the present Sudan was the setting for the disputes of the Christians and for the academies and sects of the Jews. Men and women in this region whose grandfathers were at the Council of Chalcedon lived to have Islam sweep over them. The central revolutionary concept of Islam was a radical minimalist revision of theology. In fact, part of my fascination with the role of minimalism came from visits to mosques in India, Israel, Uganda, and Kenya. In every case my guide carefully and with obvious pride brought me to see the "altar." In every case, whether in the small simple mosque in Kampala or in the gorgeous Mosque of the Rock in Jerusalem, the "altar" consisted of an absolutely blank slab of stone set in the wall nearest to Mecca. It was always shown to me without any further comment, other than a slight hint of a smile. I cannot imagine a stronger, nor more clearly communicated, statement of minimalist theology.

We wish we could have had more precise historians present at the beginning of Islam, but that is true of all major religions. It does seem clear from archaeological and literary evidence that in the centuries preceding Islam, the coast of Arabia was a crossroads for trade and ideas. There were elegant pagan temples and churches, congregations of Jews of more or less rabbinical persuasion, and poets writing in Arabic of enormous rhythmic and grammatical complexity. Tradition states that Mohammed himself was illiterate, but he certainly was aware of a highly literate civilization.

The central sacred text of Islam, the Koran, was, according to orthodox tradition, dictated to Mohammed by the Angel Gabriel in a twenty-year series of meetings that began after Mohammed's fortieth year. Mohammed in turn dictated these communications to secretaries. Just as the orthodox texts of the Old and New Testament were fixed only after a period during which many variant texts were used, so the orthodox text of the Koran was fixed several decades after the death of Mohammed, after Islam had already burst out of the Arabian Peninsula.[14] The Koran itself presupposes the reader to be familiar with at least the basic plot of the Old and New Testaments and some of the noncanonical stories associated with them. Jesus figures as a major prophet, as does Moses, and Abraham preserves his patriarchal role. The Koran deals with problems raised by these stories. For example, the Koranic chapter entitled Jonah does not mention a whale, nor does it mention Jonah by name, but rather focuses on the problem of repentance. Why did the citizens of Nineveh repent after hearing just once the words of a foreign prophet, while the people of Mecca were reluctant to listen to Mohammed, who after living among them for forty years had suddenly found a new voice and a very serious message?

Curiously enough, the Book of Jonah is now read in all synagogues as the sun is setting on Yom Kippur, the Day of Atonement. This is not because a day's fast brings whales to mind but rather just the point being made by the Koran, that the central message of the book of Jonah is that if even the heathens of Nineveh are capable of responding to a call for repentance from a stranger, why can't proper God-fearing people repent? This role for the book of Jonah in pre-Islamic time is demonstrated by the Talmudic tractate on fasting and repentance (Taanit), in which public reading of the Book of Jonah by a tearful old man is suggested as a last resort to bring people to repentance at the end of a period of public fasting proclaimed in the case of a very long drought. If rain did not follow this performance, then it had to be assumed that the drought was not due to impiety and sin but was simply part of a meteorologically abnormal year and fasting and repentance were irrelevant! In this case both the Koran and the Talmud see the same central point in Jonah. The Christian emphasis on the cetological and ressurectional aspects of the book of Jonah is quite different.

In other cases the Koran adds to Biblical stories original elements which match neither Christianity nor Judaism. For example, in the Koranic rendition of the story of Aaron making a golden calf as an idol while Moses is on the mountain, instead of Aaron explaining to the angry Moses that he was not responsible because they put the gold in the fire and the calf "popped" out, as is reported in the Torah, Aaron explains that a wandering Samaritan beguiled him into calf-making.[15]

The central Koranic message of an absolute and minimalistic monotheism and of a new beginning is clear, but the Koran itself is spare and difficult reading. Almost immediately the elaborations began, apparently satisfying the demands for human interest and legalism, leaving the theology, or absence of theology, intact. Stories of the life and times of the Prophet burgeoned. These were called *hadith*, the sayings of the Prophet. Tradition says that two centuries after the death of Mohammed there were 600,000 hadith accumulated, of which the codifier, Al Bokhari, selected 6,000 for his anthology. I don't know the criteria for selection, but at least part

of it was esthetic and dramatic. In one charming (and here abbreviated) example, a poor man came to Mohammed asking what he
should do to purify himself after having had sex with his wife
during the Fast of Ramadan. He was told to give alms, but he
excused himself by his poverty; he was then told to fast for a long
period, but he claimed physical weakness; he was then told to give
food to the poor, but he said he had no food to give. He was then
given a basket of dates to distribute to the poor, but he complained
that it seemed foolish to give the dates away when his own family
was starving, at which point Mohammed laughed so hard that his
back teeth were visible and dismissed the man without punishment.

Several centuries elapsed between Christianity's formulation as
a religious system and its acquisition of political power. This may
have permitted a focus on abstract theology and ethical theory. Just
as early political impotence permitted the Christian focus on theology, thousands of years of political impotence have permitted the
purely intellectual expansion of Jewish legalism. Islam, by contrast,
took on political power almost at its initiation. A legal system was
required immediately. This has burgeoned into the third main body
of Islamic literature, the Shira, which is still being elaborated, in
part limited in its complexity by the political realities.

In this very brief account of the three Religions of the Book we
have seen that each of them began by minimalizing the most highly
elaborated aspect of the belief system of their predecessor and then
proceeded to elaborate some other aspect of their intellectual system.
A complex theology, as found in Christianity, seems incompatible
with the kind of complex legal systems found in both Judaism and
Islam. This may derive from the dynamics of revolutionary minimalism. It may also be a logical necessity in the sense that an empty
theology can be reasonably consistent and therefore will not interfere with any legalistic deductions, while a complex theology must
contain so many inconsistencies that restrained logical derivations
become almost impossible. An axiom of modern logic, which we
will meet again, states that an internally inconsistent assertion implies all possible assertions, thereby making the process of implication itself fruitless.

Is the process of religious revolution necessarily connected with
minimalism, or have we been looking only at an oddity of Western

religion? I am not sure what "necessarily connected" means in this context, but there is some indication that the phenomenon is more general. I will consider three historically connected but discrete religious systems—Hinduism, Buddhism, and Jainism—again in thin outline since proper description of each one of these can fill an entire library and the study of the connection between them can fill another.

Revolution in the East

Buddhism appeared in the middle of the first millennium B.C., another example of the intellectual ferment that bubbled throughout the literate ancient world until it was stilled in large part by Aristotle's prize student, Alexander of Macedon, 200 years later. Buddhism grew out of a world of Hinduism. The verbalizable parts of Hinduism emerge in iconographically recognizable form from the clouds of the Neolithic long before literacy. Perhaps they shared their origins with the Greek and Zoroastrian systems, but that is not germane to our immediate concern. What is of interest is that the religions of India then, as now, were prime examples of elaboration of theology and iconography, with adjuncts of asceticsm and a strong sense of the distinction between the spiritual on the one hand and the crassly material on the other. In addition, they had the idea of transmigration and of judgment after death (found also in the Religions of the Book). In the Hindu religions the outcome of that judgment determined to which status the defendant would be reborn. To be born at all was to be sentenced to a new try at corporeal life. Only the innocent and those that had served their sentence were free of life itself. This system then could be seen in two parts, one of the iconographic and theological, the other the judgmental and ethical.

Around 400 years passed between the data assigned to the life of the Gautama Buddha and the first canonical texts of Buddhism, the Pali Canon of the Hinayana or lesser Buddhist way. As I understand it, the elementary initial form of Buddhism consisted of denying the theological part of Hinduism completely, but since Hinduism left enormous room for the imagination of the individual worshipper this may not have been revolutionary. To live, die, and be reborn were personally important, while the multitude of gods might well be either nonexistent or at least operationally insignificant. A proper and personally advantageous life, in the ultimate sense of all future lives, could be conducted without being concerned with the gods, so long as one remained concerned with moral and ethical behavior.

The standards for moral behavior set by Buddhism were not generally novel. Poverty, honesty, sexual abstinence, and disdain for material advancement were already properties of the Hindu holy men.[16] What seems to have been new was the role assigned to "illusion." Not only were the various goods of the world illusory, as was known to the Hindu holy men, but even transmigration and judgment were illusions. Nirvana, the freedom from transmigration and the state of Buddhahood, of nonbeing, were already present in each individual if illusion could be conquered and if one's perception of the world could be suitably purified. This might be the most extreme possible minimalism. Anything that could be said or felt or thought about the world was extraneous to the essence of reality.

As might have been expected, even before the Pali Canon appeared the man supposed to have originated this doctrine of the denial of all conscious ideas of gods had been made the subject of most elegant sculpture and was becoming the center of a new, highly complex, nonliteral, iconographic pantheon. Between the time of Buddha and that of the Pali Canon, Alexander of Macedon and his army had advanced to the Indus. After his death his empire lay sprawled over the eastern Mediterranean and western Asia like a jellyfish left on the sand by a high tide. As it disintegrated it left behind pockets of Hellenism. One of these became the Gandhara states of northwestern India. When the message of Buddha was received in Gandhara, it was symbolized by statues, clad in recognizable Greek robes and with the hair bound by an Alexandrian fillet. These statues quickly took on the seated posture of a Hindu god or holy man, and the same style of imagination that had generated the Greek and Hindu pantheons soon surrounded these Buddhas. They were seated with other godlike figures of many kinds—ranging from personified natural forces and demons to heroic bodhisattvas, persons whose compassion for mankind caused them to reject the transition to a state of Nirvana in order to continue the struggle on earth. Through it all the costume persisted—so that the Buddhist statues being made in Japan today often wear the fillet of the Macedonian king and are served by statues of the old Hindu gods, transformed into retainers of Buddha. Perhaps the return of elaboration was inevitable, and the only irreplaceable contribution of Hellenism was the rather trivial costume.

Often the imaginary nature of these statues was made clear by placing them on a lotus blossom to demonstrate their incorporeality, but this was also part of classical Hinduism. The weighty god of science, Ghanish, the elephant-headed son of Shiva, is also portrayed seated comfortably on a flower.

Of course Buddhism, like Christianity and Islam, produced sects and schisms, some of which, like the tantric schools of the Himalayas and Tibet, complicated things, while others, like Zen, minimalized them to the point where they could, in dilute form, become part of the southern California lifestyle. Unfortunately, we cannot dwell on these any more than we did on the similar internal shifts in the Religions of the Book.

Like many other religions, Buddhism became a state religion at several times and places, with all the complication and elaboration that entails. Curiously, despite its remarkable disclaimers, it too was capable of waging war, as in modern Sri Lanka, and of persecuting other religions. One of the victims of Buddhist persecution in the eleventh century was Jainism, which has maintained a minimalist viewpoint to an extraordinary degree, while developing a literature of enormous complexity. The origins of Jainism are even more obscure than those of Buddhism. One of several accounts assigns the founding to Mahavira, a contemporary and countryman of Buddha and, like him, of the Kshatriya or warrior and administrative caste of Hindus. I cannot tell how the behavior preached by Mahavira is related to the mythology of the Jains. I will deal with the behavior first. It is based on belief in the absolute evil of the created world and the conviction that the only way to confront it is to deny its power completely. This is a repeated theme in world religions. The list of doctrines that see the material world as in itself evil includes the Marcionite heresy in second-century Christianity, and Manichaeanism, beginning with the Persian prophet Mani's opposition to the standard Magian authorities. It continues with the Bogomil heresies of medieval Bulgaria and the Albigensians of southern France, who were mercilessly destroyed in a crusade in the thir' teenth century, through to some of the groups in the peasant rebel' lions that followed the appearance of Luther, and perhaps including the naked protests of the Canadian Dukhobors in this century.

The vehemence with which one denies creation may vary. A rather mild Manichaeanism briefly attracted Augustine. During this period he vowed to deliberately avoid procreation and creative work while striving for a state of complete inactivity. This approach would be quite acceptable in modern America. Perhaps the strongest denial can be asserted by arranging to commit suicide at the earliest possible convenience.

The Jains are one of the more extreme and persistent advocates of negating creation. While poverty as a form of sanctity occurred in both Buddhism and its parent, Hinduism, the early Jains carried it all the way. Jain poverty, for approximately the first thousand years of its history, involved not only abandoning family, home, and livelihood but all other possessions. The early Jain saints were completely nude and did not even own a begging bowl. While the doctrine of transmigration in Hinduism led to respect for animal life, and to vegetarianism, the Jains carried it further. The saints ate only vegetable and fruit soups strained through a gauze face mask to avoid involuntary consumption of insects, not for the sake of the taste or digestion but for the sake of the sanctity of existing trans' migrating life, so that it might eventually escape from the created world entirely. Even a naked Jain carried a fan of peacock feathers. This was to brush the ground before sitting down to avoid the squashing of small organisms. The ideal procedure for a saint was to starve himself to death. There are now around 14 million Jains in the world.[17]

Obvious questions are raised by this description. How could Jains have persisted, and how can they make a living? Obviously a Jain could not be a farmer, since there is no way to plow or plant and be sure you have not cut a worm in half. Not all Jains aspire to sainthood. Those that do not continue to procreate. A taste of sainthood can be taken by various kinds of ritual fasts, and it is not unusual for older people to undertake a fast to death. Also, while farming and manufacturing may be directly dangerous to life, busi-

ness and banking are not. The Jain community is generally prosperous.

Also, as might be expected in a 2,500-year history, variant schools
have arisen, some of which permit clothed saints with begging
bowls. There are even Jain statues of various gods. Jaini notes that
the making of statues by the Jains was reintroduced when they were
being accused of atheism by Buddhists around a thousand years ago!
The Jain statues are usually naked and often standing or sitting in
a curiously stiff posture. In modern Jain temples the temple itself
may be highly ornate, with representations of birds, animals, and
flowers, but the statues of the saints are stiff, simple, and repetitious,
echoing the simplicity of the Islamic altar in the ornate mosques.

Despite its austerities, Jainism seems to have a strong sense of
the relativity of truth itself. The nature of truth is said to resemble
the making of butter. If cream is placed in a jar suspended from a
tree branch and two maids pull on the jar with opposing ropes long
enough, butter will form. Truth also depends on the direction from
which things are coming and the reversal of these directions. With
this underlying disclaimer Jain mythology is free to become fantast-
ically elaborate as Campbell emphasizes, with impossibly large spans
of time measured in myriads of oceans of years, where an ocean of
years is defined as 100 million times 100 million "parvas," where
each parva is of uncountably long duration. In this enormity of
space and time live impossible persons with 126 ribs, princes studying the 72 sciences for billions of nights before passing to the next
stage in their education and so on. Once again we see minimalism
and elaboration together in the same religious scheme.

An obvious question is raised on these accounts. Is the shifting
balance between theology and legalism that is associated with the
transitions between Judaism, Christianity, and Islam or Hinduism,
Buddhism, and Jainism simply a matter of artistic taste? Or could
it perhaps be a necessary political move to make clear to everyone
concerned that a revolution has taken place? Or rather did the
undue elaboration of one of these aspects of religion over the other
in fact cause the revolution? Is excessive legalism or excessively
fantastic theology a more or less inevitable structural weakness that
invites revolution?

A fascinating hypothesis was developed by Crone and Cook in

the context of an interpretation of the early history of Islam.[18] I present my abbreviated and simplified version of their elegant and strongly documented arguments, adapting them to my interests. Using non-Koranic sources, they suggest that when the Arabs swept northward into Syria, they conquered a mixed assemblage of Greek- and Aramaic-speaking Christians, Gnostics, pagans, Jews, and Samaritans. Within at least the Christian, Gnostic, and pagan communities conflicting theological argument, combined with political and economic confusion, made it almost impossible for individuals or religious communities to develop a consistent self-image that was not denied or threatened by that of a neighbor. Islam offered these peoples the solution of a consistent self-image, at the price of shifting their language to Arabic and of reconstructing their own history to assign themselves an Arab ancestry, or at least a kinship with Arabs.

The Arabs had absorbed from the Jews the idea of a rabbinate, which in effect cut through the knots of theology but generated the legalisms that go with rabbis. The important distinction between priests and rabbis, for our purpose, is that the priest speaks from the authority of his priestly status. Rabbis having no priestly status are in one sense laity, and therefore, if they are to retain authority at all, must rationalize the legalism of their pronouncements. Speculative theology requires an authority beyond that of rabbis.

The fact that Jews have a vestigial priesthood, that Samaritans have a body of laws, and that lay interpreters of Christian religious law can be found even in the priest-centered Christian sects all represent minor modifications of the general contrast between the priests and the rabbis.

Also, a consistent and rich self-image may be aided by the reconstruction of history. People can choose ancestry to conform to a desired image.

The modern Rastafarians have created for themselves an Ethiopian ancestry; more ancestral European noblemen have been created by Americans than could possibly fit in any register of peerage; and the evidence for the existence of the Biblical patriarch Abraham, let alone his biological relation to present-day Jews or Arabs, is essentially nonexistent and would require a great deal of strengthening before it achieved the status of being dubious. The details of biological ancestry seem thoroughly unimportant compared with the need for consistency of self-image. In fact, except in cases involving laws of actual inheritance, I have the impression that undue concern for actual biological connection with remote ancestors is almost always associated with fanaticism, snobbery, or both.

When Islam reached Iran it encountered a civilization that was not fragmented in the same way as that of Syria and was primarily under control of priest-centered religions. There were Persian-speaking and Persian-writing Magians, Nestorians, and other Christians. Iranian Islam centered around the priest-like Imamate, whose authority depended on descent from Mohammed, and not rabbinical legalism. Having this priestlike legitimacy, the Shiites of Iran did not reconstruct their history into an Arab one.

In many of the American and some of the European Protestant sects, some of amazing originality, which arose out of the weakened social order and conflicting and inconsistent religious imagery that followed the Reformation, the sanctity of the priesthood and the playfully serious discourse of the rabbis were replaced by charismatic leaders whose authority was considered to come directly from God rather than from either priestly succession or rabbinical training. We will return to this type of religious thinking in Chapter 9. For another example, the revision of the status of Soviet Marxism has eliminated, or at least weakened, a caste of sanctified spokesman who had authority to decide issues ex cathedra on the basis of a Marxist ordination. During the Stalinist period, as in theocratic Russia prior to the reforms of Czar Alexander II in 1862, lawyers as we know them in the West basically had no role. Lawyers did have an importance in Russia between 1862 and the revolution of 1918, but not since then. Members of the current Congress of the Soviet Union are saying that what they need now are as many lawyers as possible! From an American standpoint this seems amazing, and the usual comment seems to be that we can set up an export business.

This examination of the role of minimalization, complication, and a curious kind of imaginative playfulness in the context of how people have structured their belief system about the world leads me to several conclusions.

Internal contradictions or incoherence in deep belief systems can be made tolerable by religious revolutions that provide new and more satisfactory doctrines. So long as incoherence is not intolerable, doctrines need not be absolutely fixed but can be modified in ways that preserve continuity without revolution. Some of these modifi-cations give every appearance of being esthetic and even playful. It is not necessary for deep belief to turn into silliness, if there is room left for intellectual changes.

An early Zionist theorist wrote an essay on the question of why— half a millennium after the invention of printing and thousands of years after the invention of the convenient bound book, or codex— parchment manuscript scrolls were still being handwritten for use in synagogues. He concluded that keeping the forms fixed was precisely what permitted meanings to change without being lost, so that meaning itself would have to be reinvented in each genera-tion.[19] This is a very modern but by no means a parochial view. The American Bill of Rights is part of our deep sense of reality. Its words have not changed for centuries, but its meaning is being continually revised. The clear distinction between those aspects of reality that are sacred and cannot be manipulated freely and those that admit of modification and even playfulness is still of major importance and will be seen in various contexts in later chapters.

How far and in what ways can doctrine be stretched without breaking? Playfulness in doctrines and in their modifications seems fairly obvious in Talmudic arguments and in the Islamic hadith. One medieval Christian example of doctrinal playfulness is the serious medieval discussion of whether or not, in an emergency, a man "could make his act of contrition into the ear of his horse."[20]

The idea of "play" itself will be clarified in the next chapter.

3

The Great Intellectual Playing Field: In Praise of Games

In the previous chapters I have outlined the thesis that organisms are sensitive to and respond to only some aspects of their environment, and that what elicits a response differs from organism to organism and even time to time in the same organism. Humans restructure their system of sensitivities and responses into a "self," which may then make more or less conscious decisions about how it will respond to the environment, or, more simply, what it will "do." This may involve shutting out all but a small subset of properties of the environment. In this chapter I will suggest that this kind of exclusion is necessary for all intellectual exercise. To simplify in this way may generate a "play" situation, although not all withdrawals from the full impact of stimuli constitute play. We are free to elaborate or minimalize—that is, to "play" with—even the most serious and intimate portions of our intellectual relation with reality, so long as a few ground rules are obeyed. Violating these rules or, conversely, taking the play with too great a measure of solemnity and pomp may prove dangerous. Playful intellectual inquiry is not an exercise in creating illusions. It is serious and may be our best access to reality. Clarification of these distressingly crypticsounding assertions will occupy much of the rest of the book and all of this chapter.

We live in a turbulent swirl of stimuli—raw information, most of which is useless and most of the remainder more distressing than helpful. Simplifying things so that we can focus on some small bit of reality is both a practical necessity and a pleasure. I share a friend with a brilliant mathematician. My friend gave the mathematician a gift of hifi earphones. These were not attached to any sound source. The gift was used with delight because the recipient required

approximately an hour of silent concentration before he could visualize the geometric structures that were the focus of his research.[1]
If he were interrupted during this period, the concentration count-
down had to begin again. He requested a second pair, so he could
leave one in the office and use the other at home. I understand he
is married.

Even nongeniuses need islands of simplification that are more or
less respected by those around them. A person behind a newspaper
expects to be immune to concentration-breaking interruptions, as
do solitary people in bathrooms, libraries, saunas, meditation trances,
and traffic. There are also places in which whole groups can reject
most of the world's flow of information—bars, synagogues and
churches, congressional and parliament buildings, sailboats, squash
courts, and the better colleges.

A generally accepted way to achieve sanctuary from the complexity of life is to "play"—to focus on a natural or artificial subset
of reality that permits some reasonable and pleasurable control and
manipulation. The idea of play is a very serious one. Johan Huizinga,
a Dutch historian, wrote two magnificent books on the relation
between play, games, and intellectual history. Both books appeared
in the wake of war, *The Waning of the Middle Ages* in 1924 and
Homo Ludens in German in Switzerland in 1944.[2] In the first book
Huizinga focuses on European court life, war, art, and poetry in the
period from around the twelfth to around the sixteenth century.
These 400 years set the stage for modern life. European languages
developed something close to their modern form during that period:
literate Anglophones today can read Shakespeare with a glossary
and Chaucer with some imagination and a dictionary, but texts from
a century before Chaucer are for scholars only. Religious paintings
of the thirteenth century may have conventional subject matter and
design, but they are representational art, while those of the eleventh
are symbolic narratives on gold leaf fields. During the same period
European boundaries approximated recognizable forms, and modern
science began.

In writing the first book Huizinga found that before he could
understand the political and social history of the period he had to
understand the elaborate playful artifice mingled with misconceptions and self-deception that characterized courtly and artistic cul-

ture. This led him to examine in the second book the idea of play itself, from the darkness of ancient times to that of the 1940s. He came to the general conclusion that in some sense all of culture is an aspect of play, suitably defined. Huizinga's books are too important for a facile summary.

Some of his assumptions are dated. Since he wrote, we have learned a great deal about animal behavior, anthropology, and evolution, or at least our opinions have changed. For example, he equates "culture" with literate "high" culture, so that he can claim that the lower classes and "savages" lack culture. Anthropological investigation in the past half century has demonstrated the usefulness of conceiving the idea of culture in a much broader sense. Also, since he equates play with the appearance of happiness, innocence, and exuberance, he uncritically assumes that all animals play. This is incompatible with our current understanding of animal behavior. Nevertheless, what I learned from Huizinga was critical to what I am trying to do in this book. Specifically, he emphasized the importance of the playing field itself, which is vital for much of my discussion.

The Biology of Dramatic Play

The literature on the role of play in health care and child development, and the possible evolutionary role of play in animals, is enormous. Here, as elsewhere, I will extract from this tangle just a few strands to weave into our central theme, much as players permit only a few simplified threads of experience to enter their games.

Reptiles, fish, salamanders, and frogs are too dignified to play, or at least do not let down their dignity like mammals. Or perhaps they are too stupid to step out of purposive approaches to undigested experience. Watching an aquarium is not the same as watching a puppy, but more like watching a busy cafeteria. Also, invertebrates, to the best of my knowledge, do not play at all.

Forty years ago I heard Sir Julian Huxley describe ducks shooting the rapids over a tidal sand bar in Iceland and then rapidly waddling back across on the sand to shoot them again. Parrots, parakeets, and canaries will ring bell toys. Lovebirds will stick colored bits of paper in among their feathers, as if ornamenting themselves.[3] I have watched swifts emerging from their towers into the courtyard of a palace in Lombardy, flying in what looked like a game of aerobatic follow-the-leader. I have also encountered a tame gull that seemed to make a game of untying shoelaces. Nevertheless, the singing and courtship rituals of birds, accepted with naive anthropomorphism by Huizinga as playful, are seen by modern biologists as deadly earnest products of evolutionary sexual selection. So I am not sure that birds play.

Perhaps something similar is also true of some mammalian play, but I feel certain that, with any reasonable definition of playing, some mammals play. Certainly mammals seem to enjoy things. During a drought in Kenya I saw elephants coming to a water hole in the early morning, alone or in groups of two to five or six, coated with red dust, ambling in dignified single file, with the larger ones leading. When they reached the crowded pool they began splashing, spraying each other, rolling over, entwining trunks with friends and relatives, like stout children at a crowded municipal swimming beach. Dogs and horses released from a building will run in exuberant circles. A classical ecology text describes hippopotamuses in water "gamboling in sheer lightness of heart."[4]

Unfortunately, the playlike behavior of mammals has become trite since the educational television channels have become glutted with beautifully filmed gamboling otters, wolf and coyote pups, and lion and tiger cubs. Nevertheless, watching cats playing with strings or balls provides a vicarious sense of gaiety, even if you don't like cats, and I don't. While balls of yarn may be rare in the wild, pouncing on things is not. Half-alive prey, including such inedible things as moles, are brought home to kittens or puppies by their mothers, and the little ones pummel and chew these grisly toys until they break. This may be thought of as a kind of educational play or even as acting-out hunting, which may or may not be dramatic play.

In human dramatic play the players assume, or are assigned, roles within which they are free to act. The script of a "play," of the sort that is seen on a stage, can be thought of as an enormously complicated set of rules, with only subtle freedoms given to the players. Conversely, if the rules are complex enough, any play can become dramatic. Mark Twain caught it beautifully, describing ten-year-old Ben Rogers coming down the street "personating" the river steamboat *Big Missouri* and all its crew.

As he drew near, he slackened speed, took the middle of the
street, leaned far over to starboard and rounded to, ponderously
and with laborious pomp and circumstance—for he . . . considered
himself to be drawing nine feet of water. He was boat and captain
and engine-bells combined, so he had to imagine himself standing
on his own hurricane-deck giving the orders and executing them:

"Stop her, sir! Ting-a-ling-ling!" The headway was almost out,
and he drew up slowly toward the sidewalk.

"Ship up to back Ting-a-ling-ling!" His arms straightened and
stiffened down his sides.

"Set her back on the starboard!" . . . His right hand, meantime,
describing steady circles—for it was representing a forty foot
wheel.[5]

Several children can cooperate to work out a spontaneous script.
Recall playing cowboys-and-Indians at the age of around eight! "You
must sneak up behind me, but I must see you and then you must
. . ." Remember? But even a bright three-year-old on a scooter can
be a train, reveling in locomotor power, narrative skill, and the
sound of his own wheels. If sufficiently young children play at some
recognizable activity, it superficially resembles the similarly ineffec-
tual "hunting" behavior of puppies or kittens, but even children
that are too young to talk embed some of their play in a scenario.
Their physical movements are part of a story. My grandson at less
than two toddled along in the wake of his three-and-a-half-year-old,
highly articulate brother, as a chair was made into a banana tree
which could be climbed, the imaginary bananas passed on to me to
be paid for with imaginary coins and then peeled and eaten with
pantomime appetite. The following day, the younger child word-
lessly dragged at his mother until she placed the chair just where it
had been yesterday. He then climbed it, picked "bananas," and
mimed the drama that he had learned from his brother.

While the ability for dramatic play does not necessarily require
the ability to use words, it does require an introspective image of
what is "correct" in the world. The chair had to be precisely in the
right place or play could not proceed. We know that a sense of
correctness expressed as an insistence on correcting deviations from
correctness is not confined to humans, but certainly also occurs in
chimpanzees, orangutans, and gorillas.[6] Therefore at least these
mammals may engage in dramatic play, although it would be very

hard to demonstrate. In short, although play at some level is found in some organisms other than humans, we do it better and in more ways than animals.

Games

Humans not only play but also play games. Games are a kind of play that involves testing our skill at playing, either against an opponent or against a standard of excellence. Any good general bookstore will have a whole section devoted to new books on particular games, their history, anthropology, and sociology, biographies of players, game rules, and strategies for winning. The rules of games, and even the vocabulary, are special for each game. Games are best taken seriously. They are a major concern of our society.

You cannot "lose" at a solitary sport (say, cross-country skiing), no matter how badly you perform. However, if you are trying to best an opponent or trying to reach some standard, then winning and losing enter and we have a game. Most games involve several competing players. But even in games (and some sports) that are played without human competitors—like solitaire or solo sailing—there is an imaginary opponent sometimes called "luck," sometimes "Nature."

Play can be unstructured, but games must have rules and restrictions aside from the limitations imposed by our physiology. Swimming, sailing, and skiing are pleasurable sports, but swimming, sailing, and skiing *races* are games as well as sports, so that the desire to excel is added to the physical pleasure of the activity.

Rules constrain how games are played. If rules can be arbitrarily changed, we are helpless; all action is futile.[7] If players break the rules, or if they insist on importing rules from off the playing field, the game is either destroyed or at least becomes a different game. In a sense the hustler, who deliberately "sets up" a large stakes bettor by choosing to lose initial games, thereby manipulating the betting odds, or the important person who insists on being treated as an important person during the game itself are parodying the game, or debasing it, as the pornographer debases art.[8]

Some games are so simple that they can be "solved" mathematically, in the sense that players can be given explicit directions for the best possible way to play, with assurance that extraneous matters such as skill and conditioning will make no difference. Games like tic-tac-toe or rock-paper-scissors, or games in which there are predeterminable odds that particular events will transpire—like dice, or like card games in which bluffing and psychological wear and tear are not factors—can be solved by mathematical game theory. To solve a game, in this context, means to choose the correct one, or the correct mixture, out of a set of alternative strategies.

A strategy is either a single response or a sequence of responses to the observed state of the game. An observed state of the game may include one's anticipation of opponents' moves and of environmental conditions. In the simple realm of the mathematical "theory of games," games can be completely characterized by their rules, by an unequivocal system for stating who has won and who has lost, by rewards for winning and perhaps also penalties for losing. Skill and strength are not relevant.[9]

Simple games that meet appropriate mathematical requirements can be shown to have optimal solutions or strategies that are sometimes specifiable in advance of the game and sometimes are contingent on moves made by one's opponent. Tic-tac-toe, for example, is best played by preventing the opponent from completing three in a row. If a game is simple enough to be completely analyzable in terms of formal game theory, it is usually too simple to be very much fun to play. If the point of a card game is to garner winnings, and if the game is completely analyzable, it is sufficient for the players to distribute the winnings and losses immediately after dealing the cards and not bother to play at all. This is often done in auction pinochle.

The most interesting of real-world games, including chess, poker, and baseball, are much too complicated for mathematical game theory. However, it has recently become possible to build computer programs to play chess and poker. Computers do not use mathematical game theory in any direct way. They rely on their vast memory for previous games, a set of strategic rules, and the fact that even the best of human players become fatigued. The first high-quality chess-playing machine won its first game against a senior chess master in 1989. I do not know the state of automatic poker players. There are Go playing machines on the market, but they are not very strong as yet. No machine has the muscle for baseball, though pitching machines are standard equipment for batting prac-tice.

The Playing Field

From my reading of *Homo Ludens* I learned to appreciate the ex-tremely important, deceptively simple fact that an absolute necessity for most games is a playing field, either real or metaphoric. Words change meaning when they enter playing fields. "Horses" appear in both chess and horse racing, but race horses cannot be expected to move like the chess horse, except in the fantasies of Lewis Carroll. All the players must agree that on the playing field only the rules and the definitions of the particular kind of game apply and that they generally stay the same between instances of the game. If the game is basketball, each basketball game must follow the same set of rules.[10]

Rules about the geometry of the field and the equipment of the play color the game. If the boxing or sumo ring, football field, tennis court, or chess board were very much larger or smaller, or a different shape, this would drastically alter how the game is best played, even if all other rules stayed the same. Golf, sailboat racing, and cross-country ski racing involve more loosely defined fields but are bounded by equally rigorous rules. Certain gambling games and word games can be played anywhere but are still within a concep-tual field.

Huizinga emphasizes that social distinctions must be left outside the playing field if the game is a proper one. In principle, on the tennis court prince and pauper meet as simply tennis players and the outcome of the game depends directly on how well they play, not on how well dressed they may be. In such sports as horse racing, football, and baseball the use of standard uniforms for all players emphasizes this point, though on informal occasions touch football can be played without uniforms, as was made stylish in Kennedy's presidential administration. Games usually permit us to struggle against an opponent in a safe, simple way, despite differences that may color all the rest of our relationships.

Of course etiquette and mores extraneous to the game do intrude, sometimes in amusing ways and sometimes in very serious ways indeed. A stale joke of TV comedy is the golf game between a boss and an employee in which there are economic penalties to the employee if he wins. A more curious and delicate example occurred in the game of Go (discussed below), during the feudal days of Japan. The normal etiquette assigns the weaker player to use black stones, the stronger white, but in poorly lit conditions it was considered advisable to let the socially superior player use the white stones, since they were more visible. Skill at socially exclusive games can be a ladder to enhanced social prestige. Conversely, forbidding access to playing fields can be a way of emphasizing class differences. For example, barring blacks from tennis, golf, swimming, and sailing clubs in America and South Africa is a notable and repulsive recent example of perversion of what is usually considered to be the proper spirit of a game.

The rules of horse racing relate to the track, starting patterns, and length of race, all of which are in the special world of the race itself. But participation in the race involves a large financial investment, just as winning involves financial gain. Money surrounds the race in many ways but is in one sense forbidden to enter the track itself. To purchase a horse is legitimate, but to purchase a race by, say, bribing the opposing jockeys is not. To get money out of the race by having the fastest ride is legitimate, but to extort money from the other participants is not. Similar remarks apply to all games.

Games may take on importance assigned by circumstance quite distinct from the designated prizes. This is most obvious in those games that emerged from a history of real conflict. Fencing grew out of dueling, which grew out of warfare. The use of button tips on the foils or the substitution of wooden swords for steel surrounds the originally bloody activity with a playing field in which the original importance has been excluded. It is still possible to imbue the outcome of a fencing match with importance, but it is importance of a different kind than attempting to actually kill one's opponent. Wagering on the outcome of sports and contests is such an assignment of importance. Even without betting, matters of pride, national dignity, or machismo may surround a game. Sometimes such strong feeling surrounds the outcome of a game that blood-letting arises from games that did not originate in war, as evidenced

by the bloody riots at soccer matches in the last ten years. Assignment of poetic importance to games is a favorite artistic conceit, for
example, the chess game between death and the knight in Bergman's
film "The Seventh Seal."

Some Properties of Good Games

Basically, all good games involve a complete, circumscribed, and
simplified model of a world, consisting of a clear playing field with
understandable and unchanging rules whose rich consequences are
then developed by the players. Within these restrictions, who wins
and who loses depends on skill, physical and psychological conditioning, luck, and other intangibles.

Good games require effort. The supposed distinction between
physical and intellectual games is overrated. Board and table games
require many hours of concentration, which demands a strong physique. The champions at such highly intellectual games as chess,
Go, and bridge must undertake a process of physical training before
major matches.[11] Conversely, even in sumo wrestling, which depends so strongly on physical bulk and in which a one-minute match
is considered to be a long one, careful intellectual game plans and
immensely rapid strategic decisions are required of the winners.
Konishki, the more than 400-pound undefeated champion in the
December 1989 sumo tournament, wasn't even a close runner-up in
the tournament of September 1989, although his weight was no less
then. He had changed his strategy between tournaments.

Games that are worth playing evoke emotions ranging from terror
to nostalgia. Downhill skiing or sledding involves exhilaration and
fear, without a strong historical element. Chess and Go and bridge
played on a serious level are part of a rich historic tradition and can
be terrifying to the players and to sophisticated observers. Kendo
and kenjitsu are very popular sports in modern Japan, where the
clash of wooden swords was a background to late morning tea
during my five months at Tsukuba University. While Japanese
sword fighting is a modern sport, in the same sense as tennis, it
alludes to memories of war with all its horrors, and at the same
time to a sense of nostalgia for samurai culture. Sailors all seem

conscious of the archaism of wind as a means of transport, and archers all seem to have heard of William Tell and Robin Hood.

Luck or randomness may or may not enter a game. The children's card game called War is decided purely on luck, without any skill being necessary. Luck has no role whatsoever in chess. Randomness may be deliberately introduced, as in shuffling cards, or nature may contribute randomness without any encouragement, as in golf.

Most yacht racing is between boats of closely matched design, but in racing of "international twelve-meter" yachts the rules permit a broad discretion in boat design. Since the optimum performance of each particular design of boat is a function of wind speed and wave pattern, the weather on the day of a race can be as important in determining the outcome as the expertise and effort of the crew and designer. Games in which randomness predominates are for children and gamblers only. Games with no randomness are a special taste.

Some games involve handicapping or class separations. Steeple-chase horses carry weights to equalize their loads; boxers are divided into weight classes. Some handicapping systems, as in golf and Go, divide players by their past performance. If players win too often, their handicap may be increased to make winning more difficult. Ideally, handicapping results in each match being an almost equal contest in terms of the likelihood of winning or losing. Handicapped games should result in close to a tie if both players are playing according to form. For each player, it becomes a contest against his own standards and the limits of his abilities. Golf can be played in this way by a lone player.

The goal of handicapping is to ensure that a more spiritual property than physique, or even than skill, decides the outcome. Some spirituality in this sense is needed for a good game. A good game permits a deep assessment of player quality. Of course, this is quality measured in the small, simplified world of the game itself. Gluttony, drunkenness, lechery, avarice, and chewing with their mouths open may or may not characterize individual players off the playing field, without really impairing their spiritual quality in the context of the game, so long as the integrity of the playing field is intact. To tamper with the spirituality of the game out of avarice is sufficient cause for termination of a player's career.

Examples of Good and Bad Games

Games can be an enormous source of pleasure, but not all games are equally pleasurable for all people. My favorite game is the Oriental board game Go or Igo. It is characterized by extremely simple rules, so that fifteen minutes are enough to permit a novice to play. To play well takes longer. I first played thirty years ago, and I feel my game really began to improve two months ago. The game is played on a 19 × 19 grid. The two players each have a bowlful of flattened stones or shells, either white or black. They take turns placing a white or a black stone on an intersection of the grid. Once a stone has been played it does not move on the board—but if it is completely surrounded by stones of the contrasting color, it has been "killed" and is removed from the board entirely. The winning player is the one who has surrounded the most enemy stones and the most empty space on the board. The game continues until both players agree nothing further would profit either of them. Another rule is that the configuration of the board must differ after each play. Handicapping, to permit interesting play between unequal players, can be done by permitting the weak player to place extra stones on the board.

In one paragraph I have given the complete rules for the game of Go, which may be the most difficult and elegant board game in the world—with academies of professional masters and a written history going back 1,300 years. The rules for chess are considerably more complicated, since there are different moves for the six kinds of pieces, a rule for capturing, plus the rules for castling and the queening of lower pieces. In addition, the goal of the game—capture of the king—is another rule. Two or three pages would certainly be enough so that players could begin to play. Handicapping in chess is more difficult because of the specific rules about different pieces. What exactly is the value of a missing horse as opposed to two pawns?

Checkers, Chinese checkers, go-moku, and Othello (also called peggoty or reversi) share with chess and Go freedom from luck—as opposed to backgammon, Monopoly, and card games, where either random events or random arrangements of objects influence the outcome of the game. Why aren't checkers and Othello considered as significant as Go and chess? Both checkers and Othello require careful and precise play, in fact often more precise than Go. (I don't know enough about chess to judge.) Some feel that they are excessively precise and predictable. Often expert opinion as to the best move to make in Go under particular circumstances is rather vague. In Go or chess the players have a choice of a variety of possible paths to follow, while in checkers failure to follow the one best path leads to defeat at the hands of a knowledgeable opponent.

Intellectual Work as a Family of Games

For good games, knowing the rules is not a problem. What is required is to choose among the rich collection of possible moves permitted by the rules so as to win. This massive simplification is common to all games and part of our pleasure in them. The participants in a game or even the spectators—the vicarious participants—have the pleasure of spending a little while in a world that they can understand reasonably well. In a similar sense the medieval cathedrals were little pieces of the city of God. Their walls kept out the mundane world. The many altars and stations of the cross, the shrines and storerooms give the sense of a small but complete city, differing from the ones on the other side of the walls by simplicity and avoidance of distraction. A good game has the same quality, but so does any completed intellectual construct.

The requirement that games all need a playing field is of far-reaching intellectual importance for several reasons. On the playing field, *only* the rules of the game apply. The fact that there are rules and playing fields makes games simplifications of life, which can call forth as much effort and attention as anything else but usually are less dangerous and frustrating than real life.[12] One primary reward for playing a game is that games are fun for the players, but usually some other reward is given to the better player, the winner. We will consider winning and losing below, but for the moment note that the reward for winning a game is usually something of value in some place other than the playing field. In that sense no game is completely separated from the world around it, despite the rules explicitly applying only to the playing field itself. In summary, games are played on playing fields within which only the rules of the game apply, but the playing fields themselves are embedded in the real unsimplified world. There is almost always tension between the inchoate confusion of the world and the simplicity and elegance of essentially all good games. It is when the players leave this game world that rewards or punishments are given for how they played. These rewards or punishments are in the currency of the real world and may be very complicated. The rewards of a game are generally not useful for the purpose of actually playing the game, although they might support the player in other contexts. This may seem obvious, but if it is taken seriously in conjunction with other insights into the playing of games, it strongly colors thinking about other issues—for example, evolution.

In the natural complex world, neither the playing field nor the rules are given to us. We initially see only the results of their operation. In a sense we are in the situation of Alice when she went through the looking glass. At first she was a tourist, a spectator, exploring a strange landscape. But then it began to become evident that the landscape was a kind of "chessboard" but with a bit of confusion in the rules. Alice then discovered that she was not a spectator, nor a player, but rather a chesspiece—a pawn. She finally joined in the spirit of the game and tried to follow the rules so as to become a queen by reaching the eighth square.

If we refuse to think about the matter, we will at best be pawns in a frightening game with unknown rules. If we try very hard to find out the rules for the "real" world, we may at best become Kafkaesque heroes. To find out something about the rules of the world, or at least to create some playing fields within which comprehensible rules might apply, is the object for all profound intellectual work. If we do this very well, we gain pleasure, and we gain at least an illusion of control, like Alice after she became a queen.

The borders between art, religion, science, and other kinds of play are not very neat. They hinge on attitudes and intentions. The doing of science involves attempts to manipulate objects and ideas in order to learn more about them, maintaining the ground rule that the objects and manipulations are publicly replicable by anyone with the right equipment and set of directions. Religion is also an attempt to grip reality but without requiring public replicability. Art releases the necessity of dealing with unique reality and permits the creation of other realities according to taste. All three are exercises that require choices about simplification and complication, since no one who has thought about it at all tries to grasp all of reality in a manipulative way.

It is easy to conceive of art as a kind of play. Conversely, most sports and games are trivial art in the sense that they are so circum-scribed by physical limitations (in most sports) and conventions and rules (in most games). The ideal bowler learns to do exactly the same thing every time with the slightest alterations for the minute differences between bowling alleys. Inept bowling may be a kind of art, but perfect bowling is not. For play to become art requires the possibility of variety.

The simplifications that make games interesting are minimalistic examples of the attempts at simplification made in all intellectual endeavor. If we define play, initially, in a simplistic and least con-troversial way as doing things for the fun of it—moving one's body and manipulating objects or ideas for reasons above and beyond the satisfying of immediate physiological or economic needs—then large parts of science, religion, and art are play. As Erasmus demonstrated that "folly," which is almost identical to what Malthus called "vice," was the source of most of what is refined and civilized, we will see that many kinds of intellectuality, and particularly the intellectual problems relating to simplicity and complexity, are part of "play," which is neither folly nor vice.

Perhaps the world is not "really" simple, but at any moment we must pretend it is, lest we be overwhelmed by its complexity. The fact that the analyses are always incomplete demotes intellectual work from the status of a game to merely the status of play. Ulti-mately, we will have to emerge from the playful world into the field which has no visible borders and in which the rules are unclear.

We will see throughout the rest of the book attempts to simplify parts of the world so that the rules can be made apparent, but the best of these attempts will not quite succeed in providing the closed reasonable world of a game. Games differ from ongoing intellectual effort in being able to permanently and arbitrarily turn off the world's complexity. Nevertheless there is a gamelike quality to most artistic, religious, and scientific activities. Ultimately, however, we are caught in a situation where we become the pawns rather than the players, or at best we try to be both at once, at which point the complexity of the world comes rushing in. Situations of this sort fill the last two chapters.

4

Three Dinner Parties

Unlike theoretical physics or esoteric religious exercises, dining is an intellectual activity available to all. It nevertheless exhibits the basic properties of other intellectual activities. Within a circum-scribed, and thereby already simplified, playing field, artistic talents of various kinds are displayed. The mundane biological nutritional process becomes the focus of art and is elaborated, simplified, or minimalized for any number of reasons, as the mundane biological acts of walking and running are turned into art in dance. By con-sidering the act of dining we can introduce, in a simple and acces-sible form, examples of essentially all that can be done in the way of simplifying or complicating. It is an especially appealing art for two reasons. First, while it certainly can be performed by highly trained masters, it is simple enough to invite amateur participation. Also, I enjoy reading about food. It provides a vicarious experience capable of much greater variety than that provided by verbal de-scriptions of sex. If you do not like to read about food, skip this chapter, but note that there is no chapter about sex.

I have largely ignored actual recipes and their history. Cookbooks are the best-selling products of the publishing industry. Hundreds appear monthly. Almost all of them have a long preface chapter on history or sociology of whatever cuisine concerns them, which ex-cuses me from repeating this accessible information. I have also ignored haute cuisine. As in the other high-fashion arts, there are critics and snobs. Styles change almost as rapidly as those of haute couture, and for the same mercenary reasons. Several years ago pasta al pesto was in but went out when sushi came in; sushi has now bowed to tapas, and nouvelle cuisine has turned old. The access of the general population to the actual eating of haute cuisine isn't much better than their chance of owning high-fashion clothing or trendy art. The situation is exacerbated because art and to some degree clothing can be enjoyed without being used or consumed. My own experience with haute cuisine is negligible. In fact my appetite vanishes at the sight of the prices, even if someone else is paying.

Haute cuisine responds to market forces so strongly that whether or not it is complex or simple, complicated or minimalized is not a simple response to an intellectual or esthetic problem. It is therefore different from what I am discussing in this book. What is relevant to our theme is dining as an intimate and generally accessible folk art.

But before food can be the focus of art it must meet some biological requirements. Biology sets limits to artistic fantasy. All organisms consume material and energy from outside their body and use it for growth, reproduction, and all of the work that goes into being alive. Eating implies the consumption of preformed organic molecules. All animals eat, but plants do not. By a rather wonderful series of biophysical and biochemical steps that are well outside our current concerns, green plants and some colored bacteria use the energy in light to split the carbon out of carbon dioxide molecules and split hydrogen out of water molecules. From these they build simple sugars. Pieces of other molecules and salts that they either manufacture or take from watery solution are combined, using the sugars as building blocks and as an energy supply, to make the molecular chains, lattices, and spirals that are used in their, and our, life processes.

Animals, most bacteria, viruses, and fungi are incompetent to perform the first crucial step of using light energy directly. There are other biochemical processes that plants perform that most animals cannot. Biochemical deficiencies vary from animal to animal, so that my dog is willing to share my meals but has no need for an external source of vitamin C and no serious interest in fruits.

Our nutritional needs are quite comparable to those of other animals. Our diet, like theirs, is partially limited by the capacities of our teeth and our digestive systems. Neither we nor lions can eat grass. We, but not lions, can eat some other kinds of leaves. We can generally eat all the meats lions can, but our teeth require that some of them be softened by cooking. Also, some morsels that delight lions appear unappetizing to me. Only in conditions of famine or extreme poverty are people willing to eat the full range of foods that their digestive machinery permits.

Most human eating is fairly casual. We feel hunger or reach the lunch hour and consume some meal or snack. There are normally bounds to what we eat, set by culture and custom. Also, *how* we eat is bound by custom. Do we eat with our hands, cutlery, or chopsticks? Do we drink soup from the bowl or tea from a saucer? Do we have napkins? How our food intake is spread over the day and what is considered food appropriate for each meal vary. Most Italians eat a light breakfast but a substantial lunch and dinner. Americans tend to have larger breakfasts, light lunches, and heavy dinners. Both may intersperse small snacks between these. Eggs may be more important in the morning than later in the day. Meat, raw or cooked vegetables, oats, rice, soup, wine, and even beer and whiskey are separately or in combination all considered as morning foods by some, midday or evening foods by others.

Almost every language distinguishes between various kinds of eating. In English, horses, cattle, sheep, and other grass eaters "graze." Deer, goats, and other leaf eaters "browse." There is also a rather mild distinction in American English between feeding and eating. Family, dogs, and children may not be the subject of such careful distinctions. To feed, when used in the context of humans, generally has a direct object. "Have you fed the children? . . . Yes, they ate." We generally would not answer "Yes, they have fed." Although we might say, "The cows have fed." Guests certainly dine, or eat; they don't "feed."

In more formal cultures these distinctions matter more. In Germanic languages the distinction between "essen" and "fressen" can be serious. When I was a child, to "fress" was a social sin, and when accused of it I had to correct my posture and get the elbows off the table and take smaller bites. Eating by the inmates of the Nazi concentration camps was meticulously referred to as "fressen"—an extra small indignity, adding a touch of horror to an initially innocent distinction.

In English the etymology of a noun often hints at class. A roof, which is viewed from the outside of a house and may require repairs, is named by an Anglo-Saxon word, while the people inside the same house are at leisure under a ceiling, an obvious French cognate. A calf has an Anglo-Saxon name so long as it must be fed and cared for, but it assumes the Norman name veal when it is eaten. The same for cow and beef, pig and pork. (Compare this to the more objective German; Kalb–Kalbfleisch, Schwein–Schweinfleisch.)

In a similar sense, "to dine" is in a different class from "to eat." While our digestive systems and teeth do not differ that much from other omnivorous mammals, only people occasionally convert eating into "dining." There is a profound distinction between filling bellies and dining. Dining involves elegance and leisure and a combination of other pleasures—as promised by Goldilocks' would-be seducer: "Goldilocks, Goldilocks, will you be mine? You'll not feed the chick-ens nor yet tend the swine, But sit on a cushion and sew a fine seam, And dine upon strawberries, sugar and cream."

Usually preparing a dinner requires greater effort than preparing a meal. People usually eat mere meals relatively rapidly, with a minimum of ceremony. They may wash, perhaps invoke God, health, or appetite before or after eating, but usually eating can't be per-mitted to interfere too much with the day's business. However, once the bare minimum requirements for nutrition have been met and either the time for a ceremony has come or time is deliberately made available, eating and preparing of foods are deliberately elaborated into an intimate art. The complications and simplifications of mere eating that transform it into dining derive from esthetics, history, literature, the graphic arts, and religion. Dining and the creation of dinners are more accessible than most other arts to amateur talent in both its creation and appreciation.

Although I remember general aspects of particular dinners for many years, dining, like unfilmed ballet, is an ephemeral experience and memories fade. I will therefore focus on three recent dinners as examples for later discussion. This may seem digressive, but it is hard not to be digressive in the context of eating. (How much productive time would have been saved if Proust had perceived the odor of an Oreo cookie instead of a madeleine?) One dinner was in Italy, one in Japan, and one in my own home. All three were parties but of very different kinds. The first was a secular party, the second was implicitly related to a metaphysical position, while the third was a ritual meal bounded by explicit liturgy and rules.

In the Italian town of Pavia, a retired professor and his wife gave a dinner. The guests brought small gifts; for example, a bottle of grappa—a white brandy—was selected by a native of Bassano de Grappa, where the best grappa is made. The table was set with antique china and a bristling armory of accessories, including five forks and four goblets at each place. The meal followed a usual Italian pattern—an antipasto of dried meats and slices of ripe melon, a pasta, a fish, several vegetables including barely cooked asparagus. With three forks now gone, a meat course of great abundance was served but was tasted rather than eaten. Drinking had gone from white to red, with the red consisting of two nineteen-year-old bottles. Sweets appeared accompanied by an effervescent wine. Finally, there were a very cold bombe and a dark brown very sweet cake. Then the grappa was opened. We were still at table at midnight.

The conversation during the dinner was primarily about food and cooking—the peculiarities of the cooking of Bologna, where the bread dough is mixed with lard before baking, the cheeses of Italy, region by region, local winemaking, the history of the dishes. The dark brown sweet cake was an invention of the seventeenth century, and the conversation then focused on why then? The cake used carmelized sugar combined with heavily boiled milk, and there was speculation about the relation of the sugar to colonial America and the relation between Italian cooking and Spanish occupation. There was also the possibility that the heavy use of sugar, at that time but not earlier, was also related to the Counter-Reformation in Spain and the Netherlands, in which indulgence in sugar took the place of less innocent sins and thoughts.[1] Did the boiled milk have any connection with the boiled milk sweets of India, the dish having originated when Indian–Italian trade was strong? Enough!

This will serve as an example of a secular, intellectual complication of the process of eating—involving pleasures that are enormously removed from filling of bellies, although the nature of Italian eating ensures that bellies are filled. The foods were the seasonal foods of Italy; the menu did not contain anything that might not have appeared in the simplest country restaurant. The simple themes had been elaborated by thoughtful choice of diners, careful preparation, the number of courses, the richness of the utensils, and the allotment of an entire evening to the meal.

The second meal was eaten outdoors seated on cushions on one of a row of smooth wooden platforms arranged immediately over a stream in the mountain village of Kibune, north of Kyoto, Japan. The stream formed the bottom of a steep-sided valley in a forest of tsuga cedars. There was a road around twenty feet above the steam, lined with restaurants, each of which owned their section of stream and their set of wooden platforms. The waitresses in informal kimonos carried the food in trays down a ladderlike flight of stairs to the diners.

The meal was served on low tables, not at all designed to accommodate my legs. There were a long series of small dishes with tumblers of cold brown wheat tea and warm napkins. Small bean-curd squares and two thin slices of raw carp were served at the same time as a vegetable plate of two ping-pong-ball sized potato spheres, a chestnut, and slivers of white radish. After a while there appeared at each place four tiny trout caught in the same stream and kept alive in a large aquarium until cooking time, fried in a thin batter and served with fried eggplant and tiny green peppers. Two or three other dishes followed, and then came small plates of pickles accompanied by filling bowls of white rice and a bowl of clear soup and finally hot green tea. Conversation was limited by language barriers, but the sound of the water filled the gap.

This was an example of a fine informal Japanese meal. Each dish was deliberately simple, with just a few ingredients, very little or no time on the stove and very little manipulation required by the cook except for careful cutting and arranging. (There were dipping sauces for some of the dishes, and I have no idea of what went into their preparation.) The setting over the stream was a large part of the pleasure, enhanced by the relation of the stream to a pre-Hiean (700 A.D.) rustic Shinto shrine. After the heat of walking to the shrine, the restaurants over the cool stream were fine to sit in, even if there had been no food. This aspect was so important that the restaurants simply closed in the cool seasons. Like other Japanese dinners, this one had at least three intentions—creating a mood, then satisfying hunger, and finally achieving a metaphysical goal.

The menu was a set of allusions to the summer season, the stream, and the location near the shrine. If the carp had been a marine fish, or the trout had come from some other stream, or the cold tea some other color than the brown of old wood, or the waitress less graceful, it would not have worked. I'm sure that each of the little dishes carried many other allusions that I didn't appreciate. Even the order of the dishes and the delay in gratifying purely physical hunger were statements about human biology and its elevation to art through the process of eating lunch. The lack of satiety that all of this left behind was assuaged by white rice, in sufficient quantity. The bland taste of the rice did not interfere with the memory of what we had eaten earlier.

All of this was a kind of craft or folk art, in the sense that no single culinary artist had created our meal. The restaurant was one in a line, not extraordinary for its class. The waitress and the chef were following their set routines, not striking out with new artistic visions, and we had simply ordered the fixed menu, contributing nothing by our menu choice. The Italian dinner was more artistic in terms of the inventiveness of the cook, but the tradition in which it was created was less formally connected with surrounding cir-cumstances than the Japanese lunch.

The Japanese lunch is similar in feeling to rare, elegant, and beautifully simple meals that are eaten out of tin plates or frying pans at the edge of a stream or a lake from which you have just pulled the fish that is the main course. Here, too, the setting and physical circumstance of the diners echo onto the food. Fresh salmon on a fishing trip celebrates a successful search for simplicity and carries much more meaning than mere simplicity by default—a salmon dinner at a seafood restaurant.

After spending some time in Durham, North Carolina, and trying hamburgers, blackened red fish, and other specialties, Tatsuo Mo-tokawa, an eminent Japanese biologist, wrote an essay contrast-ing Japanese food and the food of North Carolina from the stand-point of their metaphysics and as symbols of their respective cultures.[2] His central point is that the Japanese food preparation is anonymous. If the chef is very skillful, the diner has the sense of eating food that has not been altered by the chef's idosyncrasies. He refers to this as "no-cooking" which "let materials speak," as contrasted with what he sees as "over-cook," in which the chef "speaks." He then relates this to the contrast between Buddhism and Christianity and to different possible approaches to science. The central point is that success in "modern," that is, Western-dominated, culture hinges on a kind of overcooking. If the chef does not impress his personality on the food, he is not a chef; if the scientist has nothing startling or outlandish to say, he is not a scientist. He contrasts this with focusing on letting the world speak for itself. Echoes of his observations will reappear throughout this text.

The third meal is our annual Passover dinner or seder, the preparation of which fills our house for days in advance. The seder is a kind of domestic theater interwoven with a ritual meal. The table is set with deliberate ostentation. There are usually more guests than can fit easily around the dining room table, so that various other tables of slightly different height are joined. In a sexist distinction, men are provided with silver wine cups while women use glass. (This is not a matter of ritual prescription but just our family custom, since my supply of silver cups is limited.)

There is one silver dish loaded with four napkins interlayered with three matzoth. These are not the nice flat squares that turn up occasionally in health food stores or the supermarket. These matzoth are half burnt, of irregular shape and stiff cardboard taste and texture but renowned for the sanctity, if not skill, of the bakers. The napkin covering them tends to have waves. Precariously balanced on the napkin are six little dishes and a large leafy horseradish root from our garden. At each place is a book.

The six tiny silver dishes contain odd things—sprigs of parsley, a roasted egg in its shell, a dry and inedible roasted chicken neck, and other things that would delight an anthropologist's heart. The precise description is not needed—suffice it that none of this is casual or original. One important detail—to my left on the table, and, space permitting, to the left of the others, is a small cushion to use as an elbow rest.

The books are the standard texts at seders, to be read, sung, or mumbled aloud in their entirety before the table is deserted. It takes the place of free conversation, except that the reading can be interrupted at any time by any question about the text. The youngest competent child chants four questions about why are all these things being done (a great moment for a child to show off), and the bulk of the text is a rather convoluted reply. The details of the text do not matter for our immediate purpose.

Small bits of food, including tiny spoonfuls of crushed fruits and nuts, single leaves of parsley, single leaves of lettuce, small bites of matzoh, and several cups of wine are consumed at assigned times during the reading. The reading then pauses for the consumption of a multicourse meal prepared without flour, bread, beans, or milk. The amount of food and wine and squealing children is enormous. After the meal the reading continues, interspersed with songs and more wine, and a good time is had by all except that the children are getting sleepy and some people overeat or drink. The following night the same thing is repeated, but by this time the edge is off, the reading is a bit more rapid and the children more fidgety.

Several things are important for us. Early in the proceedings, one matzoh is broken in half and one of the halves divided among the guests. The second half of the same matzoh will be the last food consumed. This suggests the minimalist fiction that this one matzoh was the meal and all the rest were simply minor intercalations.[3] The matzoh is symbolic of poverty, so the meal is the meal of poverty. Since what is being celebrated is an exodus from slavery into freedom and prosperity, the ostentation of the setting and abundance of the courses are designed as complications to indicate new wealth from a slave's viewpoint. There is a more than coincidental echo of the feast of the wealthy freedman Trimalchio in the *Satyricon* of Petronius Arbiter.

Until around 2,000 years ago the focus of the Passover meal involved eating roast lamb, bitter herbs, and matzoh in a picnic-like atmosphere. The lamb was designated for this purpose prior to being slaughtered, and the eating of its meat was preceded by other foods to take the edge off the appetite. This early form of seder is still used by the Samaritans, a Jewish sect which has retained a priesthood that accepts the lambs as a sacrifice.

We know that the feast was transformed among the Hellenized Jews. The diners no longer had an outdoor picnic but reclined on couches, Greek style, supporting themselves on their left elbows, for which we still retain the little cushion, although the couches have long vanished. Also following the pattern of a Greek banquet, essentially as described in Plato's *Symposium*, there is a theme set by a question from a young member of the company. (Three may be pushing it a bit, but my grandson did do it very well!) The conversation, interspersed with snacks and drinks drunk at the leader's direction, then expands on the theme. The text as we now use it was written down about a thousand years ago.

The meal continued to evolve. As it is served in our house, it combines recipes brought from Europe almost a hundred years ago with modern innovations. Some things fossilize in odd ways. Green herbs figured in the Greek banquet, but they had to be bitter to recall slavery. Romaine lettuce would be ideal and is freely available in America, but we still use horseradish because in Poland no other cheap green herb was available that early in the spring. I have added lettuce in the interests of common sense and because eating horseradish in quantity can result in fainting.

The seder in my house and the lamb roast of the Samaritans and the Christian Communion are what evolutionary biologists might call homologous. They share a common ancestry, have at least to some extent corresponding parts, and serve a generally similar function. We minimalized the Biblical and Samaritan lamb to a bit of roast chicken neck on a tiny silver plate. For Christians, the lamb is a symbol of Jesus and is completely spiritual, and the entire seder has been minimalized to the bread and wine of Communion, having lost its nutritive character completely.

These three dinners, despite enormous differences, all retain the formality, the conviviality, the assignment of time, and the rich allusions and theatrical quality that are attributes of dining. They provide a basis for thinking of dining as an art form. I could have used different starting examples and seen similar things—for example, the vegetarian cuisines of India relate to the poverty vows of Hindu, Jain, and Buddhist mendicants and to the sanctity of animal life, but they can be prepared with enormous elegance and richness. In India I have been served strictly vegetarian food wrapped in edible genuine gold foil. In each case we would see a similar picture of what serious dining is about. We could also have begun with the American Thanksgiving feast, celebrating abundance, or the rich fish menus celebrating the technical fast during Lent.

While dining was often related to religion, the association is weakened at present. Dining can, and usually is, a purely secular art, as was my dinner in Pavia. Dining always transforms mere feeding into art by using food as a theatrical prop, as a wordless poem, as a way of playing on a physiological process to express meanings. Both the Italian meal and our seder celebrate prosperity and abundance and the skill of the cook. The Japanese meal focuses on minimalizing the mere animal gratification and attempting to feed higher sensibilities and becomes simple in a most complex way, by using the physical setting and method of food preparation as a commentary on the food itself.

Often dining carries echoes of middle-class life at other times and places, an imitation of a happily defunct aristocracy but without proper aristocratic insolence or faddishness. If there are no limits to the raw materials or the size of the product or the expenses of preparation, the only challenge is to achieve titillating novelty, which doesn't seem the point of dining. The fantastic menus of the Prince Regent's dinners at Brighton, with forty jellies, twenty-five kinds of fowl, and thirty of meat, and so on are well beyond the gentle poetry of normal dining, in which limited means attempt to achieve a desired effect.

The Italian dinner recalled the life of the prosperous Italian burghers before Napoleon. The Japanese meal recalled the merchants of the Edo era, enjoying gentle pleasures and staying out of the way of the dandified bullies of the Samurai class. Our seder explicitly referred to more than 2,000 years of complicated political and social history of a people that had essentially no aristocracy in the usual sense of the word.

At a good dinner a kind of communication is opened among the diners, cook, servers in a silent culture-laden language. Foreigners are at a serious disadvantage. Is the cook playing a riddle game with the diner, as in some French or Chinese cuisine in which flavors and appearances are meant to mystify? Is it a literary or historical game, as at Thanksgiving or Christmas feasts in which Longfellow or Dickens are in the background? Is it a game about poetry, as in very elegant Japanese meals? But dining is always reminiscent of poetry, in which the creator and the consumer must share a culture if proper communication and appreciation is to occur. If this communication is limited or missing, it can still be fun but on a lower level.

I hope I have shown that in one of the most intimate arts, the decisions that are made involve the same problems of modifying naive reality to satisfy taste in terms of simplicity, complexity, and minimalism as in more ostentatious activities. There is a biological component in the sense that not all possible objects are physiologically acceptable for the purpose of dining—as not all accelerations of automobiles, shapes of chairs, sizes of print, or demands on physique are physiologically acceptable. What is or is not acceptable physiologically is a result of evolutionary history. What is culturally acceptable acts as a further restriction on each dinner, so that no one cook is likely to prepare food that covers the full range of the physiologically wholesome. Often these cultural limits are not consciously noticed by the cook or the diners. Raw fish, pork, aged cheese are each highly acceptable in some cultures and simply unthinkable in others. But in each culture the range of activities available to the cook is very wide—even when there are fairly narrow proscriptions set by the nature of the occasion, location, or guests.

Dining also serves as a clear example of the simplifying role of the playing field. It introduces some of the things I will want to say about the gamelike aspect of other arts and sciences. It also hints at the simplifying effects of personal identity, both as a biological property and as molded by culture, history, and religion. These concepts will reappear in later chapters.

5

A Matter of Taste:
Minimalism and the World of Art

In the previous chapter I considered three very specific objects created by an elegant folk art that survives free of complex scholarship and legislation—the art of dining. In this chapter I briefly consider simplicity and complexity and minimalism in the context of other much more institutionalized arts. There are many elegant arts, both very private and very public, which will not be mentioned at all. Nevertheless, I hope the two chapters together will provide some sense of how simplicity and complexity function where the central issue is basically a matter of taste.

What Is Art?

We recognize the emergence of humanity as much by artifacts that evidence a deep concern exceeding the purely utilitarian as by the shape of the bones our ancestors left behind. Humans exhibit a tendency to shape, decorate, modify, or use as a substrate for decoration essentially all manipulable and significant objects. A dwelling that is merely functional, surrounded by untended ground, is a sign either of poverty, sloth, or a principled simplicity, like a meal that is only filling. And so we deal artistically with buildings, walking sticks, boomerangs, clothing, faces, even the skull shape of new babies. Painting a name on a boat's stern, or in some places painting eyes on its prow, is a proper thing to do even if we are so blasé as to no longer literally believe that boats have personalities of their own. If one is concerned about one's relation with gods, saints, or heroes, it is sometimes useful to have a symbol, statue, or picture—some object to focus on, and if that can be made by Phydias or Jacobo Della Quercia, or painted in black, red, and gold, how much more pleasant!

In the past four millennia people of vastly different styles have met, exchanging ideas and tools. While tools are fascinating to fellow craftsmen from different cultures, for most people it is the shape, finish, color, or decoration of the exotic tool that is more easily admired. Some of the statues, paintings, decorations, songs, and dances have seemed to float free of their attachment to the tools or rituals for which they were created, to become what we call art. Also, more recently, people with skills have floated free of their crafts and become not scribes, not idolmakers, not carpenters but rather artists—makers of art more or less free of other uses.

A few cheap and private art forms impose no social limits on artistic freedom. I once read of a Korean bamboo jewsharp which is of such low volume that it can only be heard by the person playing it. Certainly the virtuoso on this gentle instrument has maximum freedom, since the sound does not easily admit of criticism or competitions in skill, although it could be made more public if music composed in the privacy between the player and the instru-ment were written down and circulated among critics. The tiny engravings on little slivers of bone by Cro-Magnon artists seem almost equally private and personal, despite their imposing appear-ance when enlarged in textbooks. There are other small arts which are essentially private, so long as they can be done with cheap and common materials. However, as soon as art becomes either suffi-ciently conspicuous or sufficiently expensive or endowed with se-rious meaning, then its quality becomes a matter of public judgment. Then the choice of when the work is as good as it can be depends on several people and no longer has the immediacy of less ostenta-tious art.

The separation of artist and craftsman is still not complete, and may never be complete. Also, the freedom of an artist varies. The greater the resources involved in doing the work, the more there are likely to be social and political constraints on its performance. Expensive art is a collaboration between patron and artist. For example, architecture combines elements of craft and art in a nexus of political, financial, social, and sociological considerations. Is it an accident that the architect chosen to design the gigantic New York

State Capital campus was married to the sister of the husband of the only daughter of a senior member of the governor's family?[1] If a building emerges as artistically unsatisfactory, the architect, as artist, is free to blame the complex of powerful patrons and resource limitations within which he has worked. In fact, it is amazing that any buildings emerge as art objects. A highly cultivated autocratic patron helps, as in the construction of the Taj Mahal and the Mosque of the Rock, or at least a system in which the craftsmen and the patrons can communicate in a clear and authoritarian voice in a common language, as in the construction of the European cathedrals. It may or may not be worth suffering the authoritarians for the sake of the buildings.

In smaller and cheaper arts the separation of artistic quality and usefulness has often been mourned. A half-century ago Coomaraswamy in an influential lecture said, "To enjoy what does not correspond to any vital needs of our own and what we have not verified in our own life can only be described as an indulgence." This broadens the definition of art well beyond the esthetic ("the only test of excellence in a work of art is the measure of the artist's success in making what was intended") and at the same time reduces the significance of art for its own sake, called "fine art" ("the fine arts are the tinsel of life").[2]

Before Coomaraswamy, romantics already could sentimentalize about the passing of pure art, so Miniver Cheevey could "mourn Romance, now on the town, and Art, a vagrant." But Miniver Cheevey was fictional and drunk, and Coomaraswamy believes in simpler truths than I can accept. ("There is a 'true Philosophy' of art that can be recognized wherever it has not been forgotten that 'culture' originates in work and not in play.") Also some fine art is now a revolutionary force about which there is no need to sentimentalize.

In any case, millions of objects, from as long ago as 50,000 years or as recent as yesterday, that may have once been parts of tools, insignia of rank, or even the regalia of gods, are now called art and can be found in our museums, cathedral treasuries, homes, and monuments. In theaters, concert halls, libraries, and the air—as radio waves—are music, poetry and prose, which also are art. Surrounding art in all its forms is a heavy literature.[3] In addition to the large number of richly illustrated catalogs and histories are many, usually turgid, books of criticism and theory and philosophy. There are also books about the books on theory and philosophy. The wealth of material is excessive.[4]

Part of this richness of philosophical and scholarly analyses may relate to the fact that all except the literary arts are themselves devoid of words and, unlike most of science and technology, cannot be properly translated into words. This seems to set a challenge to talk about them endlessly, since they cannot defend themselves (although this is no longer true, as we will see). Another possible reason for the affinity of philosophers for art is that the ancient notions of causality can be so easily taught by use of artistic meta-phors and are so difficult to teach any other way. What more vivid sense can one get of material cause than an inert lump of clay and of final cause than by analogy to the goal of the creative artist when he begins his work? And how convenient that the artistlike creator god of Christianity could fit as the uncaused cause into the causal theory of the Greek philosophers!

Also, since the Greek philosophy of causality involved aspects of morality and virtue, once art was in the sights of philosophers it was peppered with these as well. As we will see, even now, with revolutionary minimalistic art, the divorce from moral issues is not yet complete. To unravel the affection of moralists and philosophers for art—as contrasted with their historically relatively low level of concern for, say, medicine and meteorology—while fascinating, is not my subject. Certain questions are, however, unavoidable, since there is so much current about art as to make simple exposition almost impossible. Art is more generally controversial than science (with the possible exception of the theory of evolution; see Chapter 8).

Apparently, American science has been tamed and brought under control, but art is still an intellectual threat.[5] What is or is not appropriate subject matter for art? How should art be supported? What are the distinctions between art, smut, self-indulgence, and propaganda? Where are the lines between incompetence and origi-nality, or between "freedom of expression," license for pornography, and a con game? Is a con game necessarily not art? Is the client the legitimate judge, or is there esoteric knowledge required for judgment of art? Is an artist an organism of special status with *prima facie* claims on public support—like a convict, baseball player, or whooping crane? How should art be taught, seen, used, controlled, paid for, or appreciated? Each of these questions has been the subject of many volumes, many hours of legislative inquiry, and much confusion.

The war on art has been conducted on at least two independent fronts. One is the subject matter of that art which has subject matter (as evidenced by concern about the words, but not the tunes, of rock music). The other is about the degree to which artists who deliberately simplify, distort, or even completely eliminate their connections with accepted nineteenth-century standards of artistic performance still maintain their status as artists. (On this front the music matters as much as the words.) In both cases what seems at issue is a workable definition of art that is evidently not obvious, although the necessity of art seems a foregone conclusion.

The subject-matter controversy in the West focuses on pornography, since blasphemy as a concern is generally déclassé. A rash of trials have attempted to decide whether paintings, films, and sculpture that portray intimate interpersonal interactions of private anatomical parts are to be protected by the constitutional safeguards of free speech. In all of these trials it is understood that art is good and is therefore to be protected, while pornography is bad, but it is not understood how either one is to be defined. I suggest that pornography is basically plodding and dull, while art is playful and interesting. Further, the producer of art is, in some sense, enjoying himself. The pornographer is almost completely bound by his subject matter and may or may not be enjoying it. This doesn't help the legal situation very much, since what is "interesting" is also undefined.

At any given time, criteria of social acceptability of interesting art vary among the arts. Mark Twain, in a cheerfully scurrilous criticism of the location of the hand of Venus in Titian's large painting in the Uffizzi of Florence, notes how what was acceptable in art and literature differed: "The foulest, the vilest, the obscenest picture the world possesses [is] Titian's Venus . . . If I ventured to describe that attitude [of her hand], there would be a fine howl . . . but Art has its privileges [over literature]."[6] In contrast with

Twain's times, we have at the moment in America almost complete license for free choice of subject matter in literature, but visual arts, including the performing arts for this purpose, have problems. Is this related to the national decline in reading scores?

Sophisticated opinion generally asserts that any subject matter at all, handled with appropriate style, is not pornography. There are alternative opinions that the definition of sophistication is as flawed as that of art. In any case, the problem of distinguishing art from pornography is the relation between subject matter and other artistic standards. But the relation between art and subject matter has undergone a long history of erosion and modification.

A sizable fraction of the arguments over art since the time of Aristotle has related to subject matter and has been tied to such terms as "beauty," "nobility," and "virtue." Only recently have the various schools of "realism" divorced subject matter from stylistic virtues. Rodin's 1864 bronze head called "Man with the Broken Nose" "was rejected by the official Salon for being offensively real- istic."[7] The long history of striving for the legitimacy of realism in literature is beautifully presented by Auerbach.[8] He makes it clear that realism is not reality but is itself a liberating and simplifying convention permitting free choice of subject matter, and greater freedom in delineating characters. For example, lower-class persons could not be taken as other than crude, or comical, or both, during most of the history of Western literature.

Pictures, statues, dramatic presentations, and verbal accounts were the only available mechanisms of preserving and transmitting intellectual subject matter until the invention, some 8,000 years ago, of mnemonic devices. Simplified pictures—of actual things (of the sort that permitted me to find my way in Japanese railroad stations and department stores despite not having a word of the language), then pictures taking on phonetic as well as visual symbolism (as in hieroglyphics and complex Chinese characters)—relieved visual art- ists of some of their responsibilities to subject matter. More effective systems of writing, cheap writing materials, and, just 500 years ago, the development of movable type printing continued this process. Visual arts by then had responsibility only for those aspects of their subjects that could not be conveniently written as words. Hamlet explains the weakness of his mother's moral situation by asking her to compare the two portrait miniatures, one of the king and one of his successor brother. Portraits were part of the luggage of emissaries arranging state weddings between strangers. Pictures, combined with a pseudo-scientific belief that the character is written in the appearance, left to the takers of pictures the transmission of subtle moral information about persons, places, and things.[9]

My use of the term "takers of pictures" for those that draw or paint portraits was a deliberate archaism. Today we use the term in the context of photography, but it entered photography from the craft of drawing, when photography usurped the craftsmen's role of providing informative pictorial representation. This opened the fascinating possibility of separating craftsmanlike focus on subject matter from a focus on art itself. I can't believe that the sudden awakening of new painting schools in the post-Romantic nineteenth century was merely coincidentally related in time to the invention of photography. Also, since photography is tied to an objective world in a way that is not true of painting, it is the photographer around whom swirls the current controversy about pornography. But if this is all reasonable enough to take seriously, the problem of defining art in a subject-matter-free context becomes necessary and important. I will suggest a very general definition and will then consider the circumstances that lead to simplification and complication of art in the context of that definition. This will lead to discussion of the extreme simplification of "minimalism."

It would have been satisfying if I could have used a biological definition, by tracking art back into the prehuman. Of course higher primates "try on" clothing and paint in front of mirrors (a very important fact discussed in Chapter 1). But they are so close to us that they hardly help. I would like more remote predecessors of art. For example, one charming candidate for an artistic animal is the satin bower bird of Australia.[10] Well prior to the return of the females from their annual migration, the male starts to construct a painted "bower." First he clears extraneous trash from an area of ground. After laying a floor of grasses and twigs on the clean ground, he builds the walls by inserting long grasses between the floor twigs, forming a bower. At one end of the alleyway is a display area on which he places blue "things," flower petals and blue streetcar tokens and porcelain cup handles. When provided with a mud of charcoal and water the bower-holding males dip bits of moss into the mess and paint the walls of the bower. Birds have been seen to use blue berries to the same purpose. By this time the females have arrived and apparently only mate with males that have created a first-class painted bower. Related birds, the parasol and palisade builders of New Guinea, plant orchids in moss which they have plastered onto tree trunks, or build twig stockades around the bases of trees, which seems to serve the same end.

Also several species of parakeets have been observed to act in a way which suggests a concern with decorative art. When investigators provided the birds with bits of colored paper, they inserted these among their feathers, perhaps in some sense improving their appearance but certainly using their feathers as a temporary storage place for nesting materials.[11] Spider crabs cover their shells with a garden of sea weeds, and other crabs plant sea anemones on their own backs or hold them in their claws like boxing gloves. In the case of these animals we can look for the evolutionary or social role of what might seem to be artistic embellishments. The crabs' sea-weed sweaters disguise their shells from the eyes of predators. The bower birds' painted bowers attract mates. Fishes' and birds' dances coordinate mating activity.

If the decorative activities of animals are associated with breeding rather than with esthetic play, perhaps what appears to be specifically human art should best be seen in an evolutionary framework. For example, in Joyce's *Portrait of the Artist as a Young Man* there is a discussion of whether the deep motivation of a museum visitor who had scrawled his name on the backside of the callipygian Venus of Praxiteles was sociobiologically or esthetically inspired, or whether the two are perhaps intertwined:

> You would not write your name in pencil across the hypotenuse of a right-angle triangle.
> —No, said Lynch, give me the hypotenuse of the Venus of Praxiteles . . . [But why this choice?]
> —The Greek, the Turk, the Chinese, the Copt and the Hottentot, said Stephen, all admire a different type of female beauty. That seems to be a maze out of which we cannot escape. I see, however, two ways out. One is this hypothesis: that every phys-

ical quality admired by men in women is in direct connexion with the manifold functions of women for the propagation of the species. It may be so. The world is even drearier than even you, Lynch, imagined. For my part I dislike that way out. It leads to eugenics rather than to esthetic . . . into a new gaudy lectureroom where MacCann, with one hand on the Origin of Species and the other hand on the new testament, tells you that you admired the great flanks of Venus because you felt that she would bear you burly offspring and admired her great breasts because you felt she would give good milk to her children and yours.

–Then MacCann is a sulphuryellow liar, said Lynch energetically.

–There is another way out said Stephen, laughing, . . . that, though the same object may not seem beautiful to all people, all people who admire a beautiful object find in it certain relations which satisfy and coincide with the stages themselves of all esthetic apprehension . . . MacAlister . . . would call my esthetic theory applied Aquinas.[12]

The love between Pygmalion and Galatea and the indecent love affair between a young man and the statue of Aphrodite in Rhodes reported in Pseudo-Lucian,[13] the pinup pictures in barracks and workshops a half century ago and the more explicit newsstand material of today, and the great appeal of other sexually titillating "art" are all subject to the same sociobiological interpretation. But ugly things, funny things, and nonfigurative things may be treated as art. Also most explicitly sexual material does not seem to satisfy the same appetite as does art. Therefore, the biological explanation for the appeal of art is simplistic. Apparently, as in the discussion of playing games, comparison with animals does not help. The creation and appreciation of art, by any serious definition, are limited to humans.

But Stephen's invocation of scholastic esthetic theory is also not very helpful, miring us back in subject matter, unless "beautiful" is understood in a very special sense—namely, of being a thing that satisfies the appetite for art. I believe such an appetite exists and will attempt to prove it, starting from introspection and then providing evidence that what I find by introspection seems general to many people, regardless of cultural and even historical differences.

Much of this volume was written in a small office in the Smith-sonian Institution, one of the greatest and most eclectic art owners in the world. Carvings from Africa and the South Pacific, jade and bronze from China, sculpture and painting from the last 500 years in Europe and America, to mention just a bit of it, were all available. When the full weight of the mundane became excessive I found that I could regain equanimity by taking a ten-minute break to see one or two bits of this treasure. Is this refreshment derived from visual arts a personal oddity?

My father was a sculptor and illustrator—active in the middle of the century. My sense of art was set by listening to him and his friends. My wife is a musician. Also, I have traveled somewhat more widely than most people. (Recall that the vast majority of humankind has never been on an airplane.) I am the product of an educational system which emphasized the virtues of cross-cultural understanding. It may, therefore, seem plausible that my relation to the art of strangers is a peculiarity of "cosmopolitanism" (in the sense used by Stalin). But museums and galleries are packed with spectators, prices are outlandish for stylish art, and being an artist has become a rather good way to make a living. Surely, it is not only the sons of artists and husbands of musicians who crowd the galleries and museums, nor are there enough relatives of actors and dancers to keep the theaters full. I conclude that an appetite for art is now reasonably prevalent.

Is this, as Coomaraswamy suggests, a decadent modern love for tinsel? I think not. There is evidence of great antiquity for cultural appreciation of more or less acculturated art objects. Carved jet and amber from Scandinavia, pottery from Greece and Italy, silk from China, engraved gems and coins from everywhere moved as cher-ished treasures over the thin spider web of prehistoric trade routes that was laid out on Eurasia. Of course nonartistic goods also traveled along the trade routes. Some objects were cherished because they represented techniques that were not locally available. Some, like ingots of metal and uncut gemstones, were raw material for local artisans or perhaps currency. But I am sure that others were cherished because they provided the kind of pleasure that I get from art today. I must conclude that my relation to art is not an oddity, but rather something I share with people over all history and all places.

How and what art is created is as intimately culture-bound as language itself. A large share of the myriad of writings on art focuses on these connections between political, social, material, and intellectual contexts and the work of the artists. The act of appreciation of art is part of what art critics teach us.[14] But the preponderance of those who enjoy exotic art know almost nothing about the esthetic theories, language, religion, politics, or even geographic location of the artists. If art is wedded to culture, how can I, and so many others, gain pleasure from art created in cultures wildly different from our own?

The degree of independence of art from its culture varies so that not all art objects are equally accessible to strangers. Some art objects, and some aspects of all art objects, cannot be accessed across cultures. For example, I can admire only the printing of a printed book in a foreign language that I cannot read, almost like admiring calligraphy of an Arabic or Chinese scribe. Translations can provide some enjoyment but of a different kind. In the first case I am enjoying abstract design; in the second I am enjoying the subject matter, cut free of the physical design of the book. I can also enjoy instrumental music which has no verbalizable subject matter.

I therefore infer that while understanding "subject matter" of art may enhance pleasure, it is also irrelevant to some degree and for some kinds of art. This is not by any means original. It has been a central belief of the most vigorous schools of art since 1800 at the latest, despite its slow general acceptance. It is, moreover, a major simplification, rendering antique and unnecessary moralistic diatribes of the sort still being produced in newspapers and legislative hearings in America.

So far, I have asserted that an appetite for art exists and that it is somehow independent of time, place, culture, and subject matter and is not an appetite for sexual titillation, nor for the culturally familiar. But is the appetite for art perhaps a hunger for novelty? At the moment, and for the past two centuries, Western art connoisseurs have cherished "originality." But originality has not been that highly prized in other times and places. When I was around ten years old, during a visit to the Metropolitan Museum in New York, my father and I spent some time looking at a Pharoah's sarcophagus. It was highly stylized, with the portrait head lying between the wings of the conventional falcon, hieroglyphics down the front, and so on. My father pointed to a line, no wider than a sixteenth of an inch, incised into the portrait's cheek, and said: "See that line? That's called freedom of expression. You will hear more about that when you grow older." Thirty years later I realized that the line represented the connections to hold the false beard. My father was wrong. More recently, I was given another reason to question the need for all but the most delicate originality to provide artistic satisfaction, when I visited the Sanjusangendo (the temple of the thirty-three bays of columns) in the Rengeo-in temple complex in Kyoto. This great hall contains 1,001 wooden statues, all of Thousand-armed Kannon. These were made from the twelfth through the thirteenth centuries by more than 200 different sculptors, and each is just different enough to be an exciting object in its own right, although they all followed the strictest of specifications, each carrying the same symbols, each covered with gold leaf, each around the same height (Fig. 2).[15]

If people's pleasure in art is not necessarily dependent on a relation to sexual desire or technical utility or subject matter or even originality, what is it based on? Is there a simple neurophysiological explanation? I know of no study on the neurophysiological or endocrinological consequences of looking at visual art, although there is some information on the physiological effects of music.[16] Using medical monitoring devices on musicians and on members of their audience, researchers have found that breathing rates, blood pressure, pulse rates, and rates of autonomic nervous discharge all can be modified by music. How much and in what ways depend on the kind of music, the sophistication and attention pattern of the subject, and whether the subject is a casual listener (or even asleep), paying attention, or performing. There are much stronger effects on the conductor than on the audience. Similar responses would probably occur in essentially any artistic "performance" worth attending—that is, in response to art "objects" that have a temporal extension. Visual arts extend in space but are usually static

Figure 2. Two standing figures of the Thousand-armed Kannon, by the sculptor Tankei (1173–1256), from the Sanjusangendo in Kyoto. The subject matter is identical, but they are significantly different as art objects.

in time. Perhaps for this reason I have not found any evidence of neurophysiological correlates among people viewing an art object.

But I do have a tentative hypothesis about what makes an art object pleasurable. It became clear to me in the Smithsonian's Air and Space Museum, the most popular museum in the world. To touch a rock from the moon, to see the planes, rockets, space ships, clothing, photographs that were actually part of the amazing program of moving into space is exciting. Each of the devices has the shape and structure that was necessary for a particular job, showing great evidence of intellect and skill but not of art in any recognizable sense. In one of the museum's galleries, however, is an exhibit of cartoons and sculpture by Rowland Emett, who has designed wildly complex, useless, and fragile imaginary machines and even built several of them.[17] The machines for the Disney movie "Chitty Chitty Bang Bang" are among the less interesting of these. There are three of his frail, feebly mobile, wonderfully iconoclastic machines in the Air and Space Museum, and from these caricatures of the technological masterpieces around them I received the signal that I get from art. This provided me one of those flashes of satisfying and untestable insight that should be distrusted at all times, but this one has refused to go away (Fig. 3).

I suggest, with all due trepidation, that what I and other people appreciate in art, and what makes it art, is the sense that any art object whatsoever could have been made differently, but deliberately was not because the artist decided that the way it was was the way it ought to be. I will therefore consider to be art objects all the things that could have been worked on somewhat more or somewhat less but that were declared to be finished when their creator decided that they were the best he could do.

This is an echo of a suggestion by my father that the character of most important sculpture and painting could be attributed in part to the necessity of performing bodily functions at intervals. This causes pauses in the work, and after such pauses the artist returns to consider his work once more and decides that it is enough. I feel ("believe" and "assert" are too strong) that some part of the pleasure I get from art is due to identifying with the pleasure of the artist at the moment the work is finished.

Figure 3. The "Exploratory Moon-Probe Lunacycle 'M.A.U.D.' [Manually Assisted Universal Deviator]" by Rowland Emett. Emett says, "The Lunacycle is intended to be soft-landed on the Moon . . . the side wheels may possibly be aligned to different points of the compass, a sure sign that the machine is poised and ready to dash off in all directions . . . The first principle in science is to invent something nice to look at and then decide what it can do . . . Should the Lunacycle fail as a space vehicle it can be marketed as a domestic carpet sweeper, to be ridden up and down the corridors and staircases of some of the more solvent stately homes."

From this viewpoint, art objects consist of those things that were declared to be in a finished state for some reason other than utility, or were intended to be that kind of thing. That last phrase is in place to accommodate such objects as Michelangelo's "Pieta Rondanini," which is now in the Museo D'Arte Anticho, Castello Sforzescco in Milan. This is the last piece that Michelangelo touched, and the work was interrupted by his death. There are other works around the world interrupted by wars, official displeasure, and so on, but these exceptions don't really bother the definition. Perhaps more interesting are works that never reach termination because of indecision on the part of the artist. For example, Saul Baiserman was a sculptor working with enormous sheets of copper which he hammered into figurative shape. I recall at the Sculptor's Guild outdoor exhibition of 1936 that Baiserman would regularly bring his hammer to the exhibition before opening hours and create an enormous din by continuing his work of hammering the piece that he was showing. It never was finished until either he died or it was no longer in his possession. It is now in the Hirschorn Museum.

"Art," as distinct from "art object," is therefore a triadic relation between artist, object, and an appreciator who may be thought of as a consumer, or observer, or patron. The artist and appreciator meet in the object. In the act of appreciation the viewer shares the last instants of the creation process with the artist.

This terrifyingly bold and simple statement seems to apply to all the things I want to include in art. These include geometric objects, like some paintings, statues, weapons, garments; purely temporal objects like songs; and all the objects that are both temporal and spatial, like dances. It permits cooking and costume design to be art but excludes some elegant and beautiful objects like intercontinental missiles and artificial hearts. It deals ambiguously with things that are mainly constructed for utility by strict rules that do not permit very much deviation but do permit some—like bridges, violins, sailboats, and automobile bodies. It also easily transcends cultural and temporal boundaries. An art object freezes the moment the creator takes his hands from it, whether that was five minutes or 50,000 years ago.

By this definition, being complicated or simple does not define art. The role of simplicity, complexity, and minimalism in art relates to the taste of the artist, and to the social context of the art itself.

Simplicity, Complexity, and the Role of Art

Most art, like most science, cannot be done without money, organization, and a clearly defined social role. Some artists do their work to worship a god, some to express themselves (which is a confusing concept), some just for the pleasure of it. Except for the last category, all art is born into a context of audience, patrons, critical opinion, and politics, in somewhat the same sense that animals are born into an ecological and evolutionary context from which they cannot neatly be disentangled.

What is desirable on the scale of simplicity or complexity depends not only on the artist but on his role in society. What will sell depends on being just far enough from the main trends of sales to appear original, but not so far as to appear outlandish or silly or irrelevant. To simplify slightly, or complicate slightly, will be more or less effective depending on the market. What is in style is a complex result of many purely intellectual and esthetic forces combined with economics and the popular concept of what art is meant to serve. If the artist is to survive economically he must provide something novel enough to be cherished, but without exceeding permissible levels of novelty. (We will touch on this again in Chapter 9.)

Perhaps a tiny bit of part of the history of European painting would demonstrate this. For this I will introduce the idea of a metaphorical transparency of art objects themselves and how this relates to the multiperson interaction between artist, object, patron, audience, censor, and critic. In musical performance two artists, the performer and the composer, are simultaneously involved.[18] Transparency in musical performance involves the performer avoiding obscuration of the music by her playing. The listener confronts the music directly, and perhaps the composer through the music. The music itself may or may not be transparent, in the sense that a transparent art object permits the observer to discern the artist. The transparency of European and American art has increased markedly.

The subject matter for medieval iconography was mainly sacred or at least homiletic. Emblems were depicted to which the observer could respond directly. For example, in the lovely Romanesque church on the isolated swampy island of Pomposa, near Ferrara, on the back wall (the wall seen when leaving the spiritual safety of the church to return to the world) the door is surrounded with vibrant, large-mouthed, red and black devils and beasts, representing the dangers one faces after passing through the door. The pictures are superb, their message clear, but the artist is invisible behind his opaque production. In fact, these figures do satisfy the appetite for art but doing so was not the primary intent of either the artist or his contemporary viewer.

With the release from the purely homiletic during the thirteenth and fourteenth centuries, the object begins to become transparent. The homiletic element is still there—the artist can rely on his audience's recognizing which is the saint normally portrayed with a hatchet stuck in his tonsure, which one carries a metal griddle, which one appears with a lion, or an eagle, or a pair of eyes growing out of a flower stalk. But the artist has begun to separate from the craftsman and become visible through the picture. In the Renaissance the hagiography becomes loose, and a conversation occurs between all parties. The object itself discourses on beauty, nobility, history, technique, all in the voice of the artist. Transparency is high; subject matter is rich. Art is very complex indeed.

But the Protestant Reformation changes the meaning of hagiography forever. Art is seen as representing not only itself and the theories about itself but the social and political situation of its creation. Rude noises interrupt the conversation. A hall attached to the Bishopric of Durham, in northern England, is lined with statues, each neatly decapitated. On one night only, soldiers of Cromwell's New Model Army were quartered there and undertook this reformation.

With the mannerists, and even the eighteenth-century classicists, the pictures retained complexity of narrative but the subject matter had become trite, the audience contracted to a mere aristocracy and the adjacent plutocracy. With some exceptions the objects had become opaque once more, except for objects like the paintings of Watteau, which turned the subject matter inside out. The elements of nobility, aristocracy, and elevation of tone curdled before the French political revolution. The art of the revolution—say, that of David—simplified both the perspective techniques and the foppery of mannerism. The object was transparent with the voice of the artist, not of the system which paid him. The artist was from then on clearly distinct from the craftsman. But the conventions of homiletic assertions, now secular sermons full of lofty sentiments, persisted.

Up to the French Revolution the lower classes aped the style of the upper. Since then it has been the reverse. Among other things, the workers in technology, whom Shakespeare called "rude mechanicals," became intellectual and esthetic leaders. Then came the camera, and there was no point in the painter's being merely a recorder. Subject matter could flow through other paths than paintings. This permitted a deeper and more interesting form of simplification than that of the market and its styles.

When subject and art became separated, some of the arts, most significantly avant garde painting, became their own subject. Painting denied the tripartite relation of artist, picture, audience—with its attendant complications of patrons, critics, and subject matter. Step by step, the focus of attention shifted. The critics and the subject matter were the first to be ignored—or at least given diminished recognition. Cezanne was among the pioneers to make paintings look like paint, rather than two-dimensional representations of three-dimensional objects. Braque removed the world further from the canvas and let internal activities on the canvas itself become the focus. At that point the playing field had become contracted, the extraneous complications of the world outside the canvas had been almost eliminated, and the communication process with the audience had become less focal. The painter was interacting with his picture; we were free to watch but not to interrupt and not to ask too many questions. The originality of this approach, like so many examples of originality, attracted disciples striving to be original in essentially the same way.

In the half century and more since the revolutionary time of Braque and the young Picasso, originality became a focus in its own right. For decades subject matter was not only passé but also considered politically reactionary. The 1950s and 1960s saw the revolutionary minimalists focusing more on the simplification process—elaborately avoiding even the painter's intimacy of communication with the picture. If there was a subject, the subject was what was meant by art itself and what should or should not distinguish a painting from a random event. Correspondingly in music, minimalism replaced the studied striving for emotional effect of the nineteenth- and early-twentieth-century prerevolutionary composers by using mechanical systems for the permutation and combination of notes. (I leave the word "revolutionary" as ambiguous as possible—asking any particular artist "Which revolution?" might result in a different answer but almost always with an attitude that you should really have known if you weren't a complete provincial, or worse, an unconscious counterrevolutionary.)

But there were no limits to minimalism and simplification and originality. One of the last complexities to go was the rectilinear canvas itself. Then chairs, brooms, single dirty or even clean words, low piles of stolen bricks, all appeared in museums, while concert halls were occupied by musical numbers consisting of a musician, an instrument, and silence, or compositions which combined mechanical combinations of things other than musical notes, such as radios and alarm clocks.

As the artists simplified their art until almost all previous understandings were expunged, they began to talk about it at great length. I recall the curious reticence of figurative artists like my father and his circle. They somehow felt that to talk about art was to act like an art critic or an art professor—both objects of scorn. But the avant garde seemed enormously capable of explaining at great length why it was they were changing or destroying things, what they were destroying, and why it was a good thing to do. In fact, some of the words are at least as fascinating as the works themselves. I will not attempt to prove that assertion here, but do not take my word for anything. Use the museums and libraries.[19] By their presence in museums and concert halls these curious objects and events demonstrated that the revolution, at least for the time being, had won.

Any revolution assumes a kind of self-organization, so that there are revolutionary company men or *apparatchiki*. The young rebel becomes the old war horse and time server, or worse yet a complete mimic of whatever he revolted against.[20] So minimalism in painting was congealed into schools, and the objects created by the rebels became subject matter for realistic representation by their students;

and the critics, like a school of remoras, abandoned the old prerevolutionary whales to attach to the aging young skins of the erstwhile sharks.

But artistic minimalism is not merely hype. At least one minimalist monument makes good use of both social and artistic predecessors and demonstrates its legitimate function of reviewing the Emperor's current wardrobe. I feel this is done beautifully by the Vietnam Veterans Memorial. This memorial contains no sign of glorification or rhetoric. It doesn't even project above the ground, so that one walks into it without notice. It is merely an obtuse angle of stone slab walls covered with names in the order of the dates they died or disappeared. It seems through its minimalism to be commenting on the apotheosis of Lincoln in his marble, self-styled temple, bordered by its unfortunate decoration of the Roman sticks bound around the handle of an axe, the "fasces," both on Lincoln's throne and on the building's entrance. The builders of the Lincoln Memorial failed to see the link between the imperial trappings and the fasces coming to life in Mussolini's Italy while the monument was being built. In retrospect, the Vietnam Memorial could silently shout its indignation.[21]

But the only revolution against minimalism is complication. The 1980s showed, and apparently the 1990s will continue to demonstrate, a return to explicit subject matter in painting, a return to "finish," and a love of complexity. Photorealism is a real movement, vying with the informativeness of the camera but retaining the personality of the artist. And of course the photographers are making statements with their work that are revolutionary in their own right, and that's where we came in with our discussion of what is art and what is pornography. Serious art frequently collides with deep beliefs, of the sort we considered in Chapter 2.

6

To Science Is Human:
Factual Pleasures

We have seen that reality is conceptualized for each individual by using a particular set of neurological and physiological structures that necessarily code or bias information. Before it can be interpreted and thought about in a conscious way, the already biased sensory information is overlain with a dense cloud of education. Education always includes explicitly or implicitly both an empirical part and a part which purports to describe the empirically inaccessible.

Several different kinds of assertions and narratives are transmitted during education. Some are strictly for artistic satisfaction and amusement. Like any other artistic creation, they cannot violate basic beliefs too severely, but within that they are basically free creations. Some are directions for getting work done. They may be embellished but cannot be substantially altered without taking cognizance of empirical restrictions. These stories are at the base of science. Finally, there are stories that are understood to represent deep aspects of reality and that are therefore sacred. "The world was created in seven days." "Bad people are punished in either this world or the next, and chew with their mouths open." "Morning is the time for prayers." These may be studied and commented on but cannot be substantially changed. They may, however, be replaced when revolutions of deep beliefs occur. Starting around a thousand years ago the slow accumulation of empirical knowledge began to accelerate, and in the last few centuries the pace of accrual of empirical knowledge has been amazing.

Not all religions focus on literal belief in a creed combined with a story. For those that did have this focus a kind of crisis occurred when new technologies and new patterns of empirical thinking intruded empirical evidence into the most central parts of some of the accepted stories. By the seventeenth century the Roman Catholic worldview had incorporated Aristotelian astronomy into Christian creed. The accepted belief was that the terrestrial world suffered permanent damage through the original sin of Adam and Eve. The word "heaven" had taken on a simultaneous astronomical and metaphysical connotation. The metaphysical heaven had been unaltered by the sin of Adam and Eve and therefore the astronomical heavens had to be perfect. Perfection involved spotlessness, stasis, and sphericity. Why the Church felt that the morality and salvation of the world depended on these particular properties I cannot say, but it is a fascinating and nontrivial question.

In any case, when Galileo reported that the face of the sun occasionally spotted like that of an adolescent, this was seen as a serious threat to religion and morality. The Copernican model of the solar system, which displaced Earth from a central location, was also a danger. If the Earth was not central, why should it have been chosen for Christ's appearance? These were the openers in a long series of empirical threats to creed. By our century the deep belief systems of most organized religious institutions had changed so as to make empirical facts much less threatening. I do not recall protests about space probes being immoral because they gave new objects to the heavens. The definition of heaven had changed. Facts can still be upsetting to creeds but only in intellectual backwaters.

Some of the empirical statements that we are taught are personally testable, but many are accepted on faith by all but a few. The substantiality of matter and wetness of water are empirically verified before the end of a baby's second year. "Matter is made of molecules which are made of atoms which are made of . . . "; "Water contains germs (except when it doesn't)"; "The Milky Way is only one of a very large number of galaxies": all are empirical statements, but very few of us have direct experience in their verification. Science is concerned with empirical assertions that may not be quite obvious to an average child.

When empirical information and technological changes became objects for intellectual consideration, it altered our sense of reality. The act of focusing on the old empirical information and deliberately attempting to generate new has become professionalized in the past 300 years. Of course Archimedes, Euclid, Pythagoras, Hero of Alexandria, Roger Bacon, and the panoply of wonderful mysterious names that file through the pages of such historians of science as Needham and Sarton were doing science for centuries earlier, but more than 90 percent of the professional scientists—supported by society to probe the empirical world—are alive now, and all the others make up the less than 10 percent that trail back into history.[1]

Simplicity within each science tends to get a technical definition. I believe that once this level of technical definition has been formulated, the problem of how simplicity is used has only a local meaning. I will discuss these meanings in the context of separate sciences, but now I want a broader context. In a general sense, how do simplicity and complexity relate to each other in studying the empirical world, rather than in the world of untrammeled intellectual creativity represented by art and religion?

The English word "science" has at least three meanings. A somewhat archaic meaning is that of a skill in performing some art or technology. Occasionally one reads of the science of playing the violin or cooking. A more current usage refers to a body of theoretical and observational statements about some circumscribed subject matter. The science of biology tells us about genes, enzymes, evolution, and, depending on where we choose to draw boundaries, agriculture, sewage treatment, marine fisheries, and medicine. Physics tells about subatomic particles, the interior of stars, automobiles, automata, and how to design screw drivers.

A third meaning, which is the one that, for the moment, will particularly concern me, is the act of formulating or discovering the elements which will constitute science in the previous sense. A student may study so that he "can enter the science of biology." It is the name of a profession, but it almost becomes a verb.

It is curious that the word science does not actually get used as a verb. The need for an appropriate verb is only partially supplied by "to investigate" or "to study," both of which are almost always transitive ("I study water fleas. I investigate how to hunt moose.") In answer to the dinner-party question, "What do you do?" I might say "I do science," where "science" is a noun, but I would like to be able to say "I science." "Science" in this sense would be an intransitive verb describing my state. "I am in the state of doing science." This would be similar to a dancer saying "I dance." Certainly the state of doing science is as strange as the state of dancing.

When a student or researcher is being most purely a scientist—rather than a technician, teacher, historian, or administrator—he is focusing on the curious interface between the empirically known and unknown. Already known facts and theories are often interesting and beautiful, but to focus on them alone would stop scientific progress dead in its tracks. To focus on the as yet unknown, without reference to the previously known, is to engage in empty speculation.

A creative scientist spends her working hours on the border between the trite, the unknown, and the unknowable. She attempts to somehow focus sufficiently on what is known so that the gaps in knowledge or the limitations in present understanding become evident. She then imagines ways in which the presently unknowable might be made knowable, and tries to clear her head of preconceptions and misconceptions that prevent her seeing the world.

Sometimes we can see a clear hole in the fabric of knowledge. After it had been made clear by classical genetics that genes behaved as particles, and then that genes were associated with chromosomes, it was obvious that there would someday be a relatively complete theory or model of a gene in chemical terms. The absence of such a model was a self-evident gap in knowledge, which was eventually filled by the Watson–Crick DNA model. Their autobiographical accounts make it clear they did not simply stumble onto their model without expecting to find it.[2] It was what was needed to fill the gap, and if they had not found it, someone else would have. The development of this model permitted new questions to be asked about genetics and development, which generated what is now an ongoing torrent of research.

Sometimes the hole is very big. Study of the problem of "self" began much before the study of the gene but is not yet solved. We know that we have a sense of a subjective self. We also know a great deal about the anatomy and physiology of neurons. How can our knowledge of neurons and brain structure explain the sense of a subjective self? Is there a way of asking that question which will transform it so that the kind of speculation about the self that I presented in Chapters 1 and 2 is replaceable by a more formal and testable theory? This may also be considered as a hole in knowledge, but with such poorly defined edges that it might equally well be a question about the borders of science, rather than a gap within a science. In both cases, however, the business of science is to proceed from the known into the unknown, from phenomena that have been tamed and coordinated into an empirical system to phenomena that seem to be real but as yet make relatively little sense.

To train new young scientists, we teach what is of relatively little scientific interest, that is, the already known, as a pointer into an area about which we can say nothing. Young scientists are being trained to attempt to break into the unknown by seeing the known in some new way, so that they may learn things that their teachers do not now know. To be trained to become a scientist is to proceed through a kind of initiation, after which the student may be trusted to behave, at least during working hours, the way a scientist ought to behave—to be a scientist, that is, to do, rather than study, teach, or memorize, science.

A similar idea of initiation of neophytes is found repeatedly in poetic and religious thinking. The mandalas of Mahayana Buddhism are used as educational devices in initiations. Mandalas are objects for contemplation which aid in understanding doctrine. What I gather from the texts is that a particular teacher, having achieved a particular kind of insight, may use a particular mandala as part of a pedagogic program to permit a student or disciple to reach enlight-enment. Elegant and complete books of mandalas are now for the first time appearing in print, because Tibetan Buddhists, now exiled to India, are publishing them to avoid their being lost or forgotten.[3] Several things are evident from turning the pages of these books.

It is most obvious that mandalas are deliberately made too com-plex to be absorbed at one look. There is the entire object, and it can, with appropriate squinting, be seen as a blob of color with a shape; but it is impossible to not see that inside the blob are whorls within whorls, patterns changing into not only other patterns but other kinds of patterns. All that I have seen have some closed geometric figure near their outer edge. There is a clear sense that this outermost closed figure, which may be a rectangle, a circle, octagon or triangle, is a kind of wall or barrier. Within the outermost closed figure are others, perhaps one circular wall in the center of the outermost wall with other circles arranged around it or perhaps a square within a circle, and further closed figures within these. Some of the circles are simply colored bands while others contain figures of gods, either within the circle or making up the circle. The number of gods represented in a mandala ranges from one to thousands, and the representations may combine human, animal, and purely fantastic forms. Each god may be alone; or male and female manifestations of the god, sexually joined to each other, may be shown. In the most beautiful edition of this sort that I have seen, the text associated with each mandala gives the name of its originator and a list of the names and sometimes some of the properties of each of the many gods that appear (Fig. 4).

Figure 4. "Mandalas" from three worlds. (a) A Tibetan Buddhist mandala (above). (b) A German medal of 1936 honoring the discoverer of a system of knowledge. (c) Part of a Jewish amulet for the protection of infants. In each case the viewer must pass through outer barriers to reach some essential central information, understanding, or relationship.

What seems apparent is that the mandala represents the unknown and perhaps the unknowable that the initiate, by suitable
learning, exercises, and guidance, may come to understand. The
initiate is to pass through the closed figures, one after the other,
despite the absence of evident open gates, until at the center is a
god with which he can'identify and, according to some teachers,
"become." The exercise is to contemplate the complications of the
mandala as a metaphor for the appearances of reality, but by con-
centrating on it with suitable intensity to ultimately see the illusory
nature of both the metaphor and the mandala itself as it dissolves
into a kind of cosmic minimalism. T. S. Eliot expressed a similar
idea in "East Coker":

Figure 4b

> To arrive where your are, to get from where your are not,
> You must go by a way wherein there is no ecstasy.
> In order to arrive at what you do not know—
> You must go by a way which is the way of ignorance.[4]

Some natural objects look like mandalas. A snowflake, the shell of a diatom, or a protozoan each at first glance looks like a finely made lace with a hierarchy of patterns within it. So also does the flower of the weed which is called wild carrot in its English home but got the fine name of Queen Anne's lace when it came to America. Scientific theories may be graphically presented in diagrams reminiscent of mandalas. The diagrams of spheres and epicycles of Copernicus look like kabbalistic diagrams of the qualities of God. On suitable contemplation, with the help of collateral observations and experiments, the complexity of each of these natural objects produces insight into simple mechanisms and rules by which they are generated.

How similar or dissimilar is the initiation of the disciples of a holy man to the initiation of young scientists? The focal distinction

Figure 4c

between the diagrammatic objects of scientific contemplation and those of religious contemplation may be that the point of the tantric or kabbalistic mandala is to bring the initiate into relationship with the private world of his teacher, while the point of scientific initiation is to relate to a public world, ideally unobscured by the personalities of the investigators. To the naive observer the world has the impregnability to understanding that the mandalas present to the noninitiate. To become a scientist is to undertake the exercise of passing through the opaque walls of knowledge, to find new walls, or new understanding. Scientific initiation is not into an esoteric doctrine. The objects of scientific contemplation are supposedly provided by nature in a much more public way. Also, while doing science is pleasurable, being initiated into science is not generally expected to involve the deep psychic changes that are often expected to accompany religious initiation.

What emerges from the scientific exercises is a swirl of laws, rules, and mechanisms that generate not a mystic insight or enlightenment but rather the everyday world of the two-year-old, enormously enriched but forever preserving the fascination that one feels in two-year-olds, that around the next corner, behind the next door, a whole new world of exciting things is waiting.

The broad lacework on the wild carrot is not a single flower at all but rather a spray of tiny florets—what botanists call an umbel. Each component floret is a simple ring of white petals surrounding reproductive parts—all but one in the center, whose petals are velvety purple brown verging on black. In a recent discussion an eminent evolutionary biologist despaired of ever knowing the evolutionary meaning of the single black floret in the carrot flower, and by simple observation it cannot be known. However, if the universe of contemplation is enlarged from the flower itself to its ecological context, the black floret leads to new levels of meaning. If some carrot flowers have their black floret removed and others do not, and we sit and watch the assemblage of flowers on a quiet sunny day, we will observe that flies visit the intact flowers more frequently than the mutilated ones. The black spot on the white flower seems to act as a decoy fly.[5] But this obviously leads to new contemplative questions. What is the role of the fly in the life of the carrot? How did the carrot come to build a decoy? What are the developmental or biochemical mechanisms and their genetic foundations for making just one black floret in a sea of white? We would have to incorporate genetic and evolutionary information into the discussion, and the behavior of insects would be part of it. But that would lead to the study of behavior in general, and so on.

As a pictorial mandala in space, the carrot flower is a curious failure. Its gross anatomical structure is visible after only a brief contemplation, but as a mandala in a broader context it is almost endless. It is easy to imagine how one could spend a lifetime in analyzing the implications of a wild carrot flower, and not waste time in the process. What emerges would be information of interest to the gardener, evolutionist, biochemist, and perhaps to a better kind of poet. We see that the tantric painting of a mandala, which is to be contemplated while seeking a minimalist metaphysical void, is almost the reverse of what can occur in contemplating the empirical world. In the tantric exercise perhaps some final state of enlightenment may be reached. In the scientific enterprise there is no such finality. Any thread that we pick up will, if we follow it carefully, lead us into a wonderful, and perhaps endless, tangle of new questions.

We will see a bit more of snowflakes and shells as we go on, and find that they connect with a different body of rules than do carrot flowers, but it is already obvious that one of the first simplifications of science must be the choice of the general bounds of a body of investigations. For most purposes the carrot flower must be kept carefully distinct from the Tibetan mandala or the diatom shell. We might hope to simplify the world by considering them together but, unlike artistic or religious simplification, which is to a large degree a matter of taste or style, simplifications in science can only be along certain natural lines. There really is a set of natural divisions of thought, and failure to take them into account results in missing the paths to insight. Conversely, the proper initial definition of a problem not only will permit it to be solved eventually, but the solution will generate new and even more fascinating problems.

Technology and the Beginnings of Science

The first step in beginning a modern scientific investigation is to define the boundaries of the problem. Doubts about projected work increase as the boundaries are extended. This is so widely accepted that anyone who claims to understand everything, or worse still has a new theory which explains everything, is almost casually dismissed as a crank. This was not always the case. When little enough was known, it was possible to know everything, but the knowledge had little depth. How did our present attitude toward the empirical world as an object of study originate?

Before science, and free of many of its philosophical and conceptual problems, was technology, the development of methods for performing specific useful tasks. The shaping of pieces of wood and stone for specific purposes, the peeling of fibers from specific plants to make twine, baskets, and rope, the carding and weaving of wool and hair to make clothing, the care of domestic animals and plants and the preparation and preservation of meat, milk, and harvested fruits, seeds, and leaves all were developed in prehistory. The use of fire, tools, agriculture, the domestication and harnessing of animals, ploughing, pottery, spinning, and dyeing, and the use of pigments and metals, ships and wheeled carts all were accomplished in prehistoric time. Inventions such as the arch in building, and such machines as the windlass, pulley, lever, rotary quern, bow drill and lathe, and numbers were also prehistoric, as were the empirical foundations of anatomy and of medicine.[6]

When pebbles bake under a bonfire, the color change is permanent. In the back of a cave near Marseilles five undisturbed hearths recognized by their reddened stones have been found. Wildfire does not enter dark stony caves. This first clear evidence of the control of fire is around a million years old—older than *Homo sapiens*. Did more advanced technology wait on the evolution of more profound intellects, or did the beginnings of technology provide the evolutionary impetus for the development of intellect? The dim beginnings of human technology accompanied the beginnings of humanity itself, and we will not attempt to explore them. It seems clear that early hunts, perhaps using fire, stones, and coordinated groups of beaters to drive the game and organized teams of slaughterers and butchers, were accompanied by the exchanges of information and most probably by stories about past hunts and plans for future hunts. The same must have been true for the gathering of edible fruits, leaves, seeds, and small animal foods like insects, frogs, lizards, and rodents.

Some technologies have changed very little and are still of value. My father, daughter, and one of my sons are professional experts in ancient technologies. It took my daughter many years to become an expert handweaver, and my son, after fifteen years of practice, now makes fine violins. Among my son's most cherished possessions is a recently published set of full-sized plans of the violins of Stradivar-ius, made 300 years ago, and my father illustrated his book on sculpture techniques with photographs of the tools of Egyptian sculptors who died more than 2,000 years ago.[7] The techniques of the handweaver and sculptor used today would be recognizable by our ancestors of 30,000 years ago.

The distinction between technology and science is not absolutely clear, but it relates to the goals of their expert practitioners and the uses to which they are put. Technology has the goal of practical utility, while science has the goal of constructing an empirically demonstr-able narrative about the world. Individual tanners can make leather from hides without having a theory of chemistry, while individual chemists may understand what is involved in the transition between a fresh hide and a sheet of leather without being able to substitute as tanners. Ideally, science can aid technology, and as a historical matter technology has provided challenges for science. Technology began in the prehistoric past, but science in anything that even approaches its modern sense is not more than 3,000 years old at most. Intellectual inquiry into nature for its own sake began with the Greeks, but the science of the Greeks was so intellectual that it failed to interact strongly with technology and therefore could not advance as rapidly as the science of the last 400 years. (The word "therefore" in the preceding sentence is loaded with assump-tions and arguments and is not to be taken lightly.)[8]

New technologies, based on science, are being created at an accelerating rate. The gathering, manipulation, and dissemination of permanently recorded information began 5,000 to 10,000 years ago. It was accelerated by the spread of printing 500 years ago and has made startling advances in just the last century and a half, when electricity began to be used to handle information. Morse's telegraph permitted a short message to be sent from Maine to Texas in under a minute, prompting Thoreau's question, "What if Maine has nothing to communicate to Texas?" New superspeed communication systems have now relegated the telegraph to its place as a sixth-grade science project but leave Thoreau's question unanswered.

My other son is a physician, an occupation in which recent changes have been remarkably rapid. While medicine is among the most ancient and revered technologies, a case can be made for the assertion that medical ideas and techniques that are more than a century old are of mostly historical interest. Earlier medical technology permitted some development of dietetics, midwifery, boil lancing, bone setting, and amputations, some kinds of psychiatry, and some aspects of public health. These were developed in the dim past on the basis of essentially no science at all. Scientifically based medical technology started only with the study of human anatomy during the last half millennium and began to be really effective only with the development of aseptic techniques and anesthesia in the nineteenth century, and with synthetic drugs and antibiotics within the last fifty years. During all but its most recent history the ministrations of highly trained doctors may have caused more deaths than cures, an example of a technology sustained by hope rather than results.

Metallic ores were not suddenly created 5,000 years ago when the period of rapid technological changes began, nor was the evolutionary state of the horse or the geometry of the wheel or the neurology of humans different 20,000 years ago from what they are today. There is no obvious reason why the Industrial Revolution should not have occurred 10,000 years earlier or later. We do, however, have a broad sense that things suddenly began to move and change. To speculate on how or why the acceleration of change began is unclear. It almost certainly did not have a single date or place of beginning.

The earliest technologies for which we have direct evidence involve a kit of simple stone tools. Stone heads for clubs and stones used as projectiles and stone choppers and scrapers were already an obvious aid in catching and preparing food. Useful objects made of wood and straw that coexisted with the earliest stone tools have mainly rotted away. From the existence of tiny stone flakes and curved bits of bone, sickles are inferred to have been made 20,000 an objective skin-color change, while the other reports nothing of the kind.

If two different people observe the same animal, perhaps a baby pig, their personal response may differ. One may be revolted, while the other may develop an appetite, but if both of them are now told to shoot off a blank cartridge in a pistol, the pig will probably run away. Conceivably a scientifically oriented director might be interested in the responses of the people, but if so he will have a much more complex problem on his hands than if he focuses on the response of the pig. Particularly, if the point of this exercise was to explain science to the two actors, only by focusing on the pistol, the noise, and the pig would the director be likely to get across the idea that science is the study of those aspects of the world about which people can agree, even if they differ enormously in many other ways. Removing clothing is not a neutral activity, and perhaps no activities are really neutral, but for all practical purposes simple activities such as laying a ruler against another stick and measuring its length in inches or looking at a bird in a tree and ringing a bell when it flies away should give approximately the same result no matter who does it.

Fortunately, the general ground rules of science are reasonably well understood, so that important scientific disagreements, unlike religious or political disagreements, eventually get settled to everyone's satisfaction. Basically, the test for scientific validity consists of having several people agree on what has occurred under reproducible circumstances, and scientists have agreed to confine all professional discussions to what has been, or in principle could be, agreed upon in this way. Without this immensely simplifying restriction, science would founder. It is not a disgrace to a scientist to have been wrong, if he was wrong in the correct way.

If there is no agreement after further discussion and observations, then either the subject matter must be eliminated from science or the analysis of the situation is incorrect or incomplete in some subtle way. Perhaps the participants have personal commitments that interfere with their objectivity. Often it is discovered that disagreements are due to the same words' having different meanings to different people. That statement will be returned to. It is more complicated than it appears to be.

Elementary assertions of neutral and reasonably objective facts

years ago and more. Sickles infer harvesting of plants, from which is inferred possible domestication of plants. A domesticated plant or a captive dog or pig must become a focus of attention and speculation. Some of the speculation would relate to the very empirical problems of how and when to plant, how to store produce, prepare the fields, cook the harvest. How long was it between the corraling of horses as meat and their use for transport?

It has been suggested that perhaps excessively successful hunting some tens of thousands of years ago eradicated many of the large mammals that roamed the pre-Pleistocene world, making agriculture necessary.[9] But a planted field, waiting for harvest, is a hostage to fortune; animals and humans may steal the farmer's labor. A fence can keep out animals, but protection against humans is more difficult, perhaps requiring a tax payment to a local bully—a petty lord or king—who will keep other bullies at bay. Agriculture in areas that required irrigation may have been the stimulus for the origination of powerful central governments, which at least in Egypt, Mesopotamia, China, and the Andes centered on massive waterworks with their associated administrative and occupational specializations. These certainly stimulated the development of new technologies, but Needham and his coworkers have demonstrated that the connection between technology and science is not obvious.[10] They have shown that many technological developments that appeared in Europe in the twelfth and thirteenth centuries had already been developed in China as much as a thousand years earlier. Movable type, gunpowder, windmills, and even clocks were among these. Even the simple horsecollar, which permitted a horse to pull hard without cutting off its own air supply, began in China but did not appear in Europe until after the tenth century. With a collar, a horse becomes a beast of traction, rather than a one-man riding vehicle. In the absence of some such gear, horses could not effectively pull anything much larger than a two-man chariot.

Despite China's technological richness, deliberate intellectual inquiry seeking natural laws and regularities for their own sake did not become a self-accelerating intellectual force in China, but rather really got started in the unsanitary and curiously barbaric Europe of the thirteenth century. Historians speculate on various possible causes for this. One possibility they suggest is that the religious structure of the West centered on a god that had the properties of a ruler who could make laws for natural behavior in much the same way that secular rulers make laws for human behavior. In the West, therefore, it may have been easier to believe that laws of nature can be discovered. The Chinese image of deity was less anthropomorphic. They also point out that for almost all of its recorded history the path to personal advancement in China was through the civil service, which emphasized literary accomplishments and failed to reward entrepreneurial adventures of a practical kind. In one sense this was the obverse of the Western situation.

In China technology flourished but theory was uninteresting, while in Greece speculation was rampant but technology was of less interest. This difference may have been due to the relative prevalence of slavery in Europe as compared with China; slaves may have made labor-saving devices less important to the West. Needham suggests that this might explain, for example, the West's failure to develop a proper horse harness, since human draft labor was there in abundance.

In the medieval city-states of Europe, with their ambitious warlords, practical and intellectual innovations were perhaps better rewarded. That is, many historians have suggested, without being too dogmatic about it, that science, capitalism, and a Protestant conception of god as a legalistic ruler fed on one another in Europe in a way that was impossible in the larger, better organized, intractable Chinese empire.

Aside from the problems of economic support for scientists, were there purely intellectual barriers that prevented science from developing earlier? Perhaps science is such a simple thing that it is difficult to do except under very special circumstances. To see the world with the intellectual equipment of a grown-up and the innocence of a two-year-old is not easy. As we mature we have a sense of the fitness of things, how things must naturally be. This can be enough to stifle any hope of scientific success. Aristotle had a clear sense that it is only natural for heavy objects to be below light ones, that circles are naturally more perfect than noncircular paths of objects, and that objects tend to stay at rest unless we push them. These reasonable but incorrect assertions were enough to impede the development of physics and astronomy for a thousand years.

For centuries moral lessons were assumed to be found in the natural world, and in fact were thought of as a possible key to

understanding the natural world. The natural histor
teenth century and even later in Europe was thought t
to preachers as to compounders of medicines.[11] To real
the ways of humanity are not necessarily the ways of
creation was very difficult to do. Even now natur
seen to be following enormously general laws tha
morality and political opinion. Christian ways of
science and Marxist ways of thinking about scienc
ously promulgated by modern academicians. Abel
century attempted to separate intellectual inquiry f
assumptions of the church and the government. S
when this is extended even further, and we sep
we can) even our personal preconceptions from s

The natural world need not be logical in any
ence does not consist of imposing our reason on th
reducing our preconceptions to the point that th
logic on us. This is very difficult indeed, involv
of our ego while maintaining our full powers
receptivity. The capacity to perform this feat i
science, as a verb, attempts to foster in the stud
completely. But it is possible to be trained to s
of the world with reasonable clarity. The firs
what must be ignored, what simplifications
one can even start.

To grasp the basic idea, consider an in
absurd. There are two persons, the actors, c
different worldviews. Add a third person
opinions to act as a kind of director. Th
actors what motions to go through. These
ulating or describing actual objects and
assumption asserts that if the same, or v
are undertaken, part of the description o
agreed upon by all concerned, regardles
they might have. Obviously, neither al
scriptions will work equally well for
might, for example, ask the two peopl
clothing and ask them to indicate cha
If the clothing and the people are app
people might report an acute sense of

ignore most of reality in the same way that even a color photograph is only a representation of one bit of reality while saying nothing about the rest. A large collection of pictures or a sequence of many scientific statements presents a greater fraction of reality but obviously not all of it. Science as an institution makes a claim that goes beyond that of photography. Namely, as science proceeds it will asymptotically approach a value-free, objectively verifiable representation of all of reality that can be objectively verified. That the pig ran away at the sound of the shot is an easily verifiable assertion. But the shot itself, the motivations of the participants, the hormonal changes in the pig, the effect of the pigs hooves on the ground, and the physics of movement of the pigs bones and the biochemistry of its muscles are all aspects of the total situation that can be scientifically discussed. The emotions generated by the entire procedure in the various human participants can also be discussed, but not as easily.

Not all scientific experiments are as silly as shooting blanks at a pig, but it really is hard to see any practical point to a surprisingly large share of investigations. In 1949 I gave my first presentation at a national scientific congress. At that meeting I was fascinated by one research title, "On the response of the alligator to the note of E on the French Horn." My own paper was on population growth in the water flea, and it wasn't until I just wrote that sentence that I realized that my subject for investigation might sound just as strange. I know the implications of my own work for general ecological theory, including problems of fisheries management and other practical things. I can imagine that the paper on alligator responses also had broader implications. Nevertheless, the raw titles do sound odd. Senator Proxmire of Wisconsin gained publicity as a shallow thinker of historic proportions by pouncing on titles of federally funded research projects and nominating them for what he called "The Golden Fleece Award." The name of a research program has no more to do with its quality and importance than does the name of a painting.

The Pleasures of Sciencing

The usual rationale for science funding given to newspapers and congressional appropriation committees is that enrichment of science leads to technological advance. That is a reason for societies to provide money to support science, but it is not a reason for young people to dedicate their lives to it. While my work might have had some practical applications, I was really doing it for more personal reasons. Compared with many occupations, being a scientist is generally not a lucrative business. Most young scientists choose their careers because the act of doing science is remarkably pleasant and, sometimes, exciting and adventurous. Even if the rest of the world is too complicated to trust or understand, science usually retains its character as a clean intellectual game.

I was a child in the East Bronx during the Depression of the 1930s. My world was thick with contradictions. During the week, I was being taught to read from a textbook that focused on a completely unreal world containing Billy, Jane, Beverly, and their little dog Spot who live in a picket-fenced house in a suburb like the one I now inhabit but whose reality was inconceivable to me then. What really existed in the neighborhood was never mentioned in school. Crotona Park is now a very dangerous place, but then on any sunny Saturday or Sunday, between the children playing punchball, stickball, pottsie, and other now-forgotten games, all watched by mothers and grandmothers, it was filled with male advocates of capitalism, Communism (with all its bewildering branches and relatives, from Syndicalist Anarchists through to Trotskyites and Stalinists), and of causes ranging from Labor Zionism, the various socialist programs such as those of the Bund and Workman's Circle, through to advocates of religious piety, atheism, or emigration to the Jewish Autonomous Socialist Region of Biorbidjhan in eastern Siberia. I now recall it along with the line in Don Alhambra's song in Gilbert and Sullivan's "The Gondoliers": "Two party leaders in each street, defending with no little heat their various opinions."

Afloat in this sea of untestable, passionate, but often highly intellectual assertions I found scientific facts to be solid, real, and inviting. When I was six my grandfather showed me how magnets moved iron filings on cardboard into patterns and when I was eight he made a Cartesian diver out of a cork, lipstick case, and a medicine bottle covered with a rubber dam and told me about Archimedes in his bathtub.[12] Then he moved away and I waited until seventh grade when the first school course in science began. I clearly remember shouting down from a second-story window that I couldn't come out and play—I was doing science homework—putting as much volume as I could manage into the word "science."

When I was eleven years old and had permission to wander over the city by subway, the Natural History Museum and the Bronx Zoo were sources of deep pleasure. In those days the animals were displayed in rather simple cages with a minimum of didactic additions. The library was there to help me understand them after I had seen them. The museum was a pioneer in the presentation of dioramas containing lifelike taxidermic masterpieces in a simulated three-dimensional setting, but it also displayed endless rows of bird skins, shells, and bones which gave me the same sense of primary experience that I got from the zoo. I also had a toy microscope which permitted me to see rotifers and ciliates in stinking infusions of lettuce in tap water. Aquaria of fishes, salamanders, turtles that came by mail in return for Ralston cereal boxtops, a pink-eyed white rabbit, a canary, and even a horned toad sent from Oklahoma by an uncle who was a cavalry sargeant at Fort Sill all shared my bedroom in a four-room apartment.

The kind of delight I felt in the process of seeing these things for myself, watching them feed and reproduce and cleaning up after them was presented nicely by Sir Thomas Browne, living in the passionately opinionated world of seventeenth-century England:

> The World was made to be inhabited by Beasts, but studied and contemplated by Man: 'tis the Debt of our Reason we owe unto God, and the homage we pay for not being Beasts. Without this the World is as though it had not been, or as it was before the sixth day, when as yet there was not a Creature that could conceive or say there was a World. The Wisdom of God receives small honor from those vulgar Heads that rudely stare about, and with a gross rusticity admire His works: those highly magnify Him, whose judicious inquiry into His Acts, and deliberate research into His Creatures, return the duty of a devout and learned admiration.[13]

This sense of delight is frozen out of most current scientific prose because of restrictions in vocabulary which seem necessary to ensure precision, shortages of space in scientific journals, and also because science for many has become a rather stuffy job which you do if you can't find more lucrative work. But when modern science was beginning, the sheer fun of it sometimes broke through. Harvey's original description of the cardiac and pulmonary circulation contains the following:

> I have seen the first rudiments of the chick as a little cloud in the hen's egg about the fourth or fifth day of incubation, with the shell removed and the egg placed in clear warm water. In the center of the cloud there was a throbbing point of blood, so trifling that it disappeared on contraction and was lost to sight, while on relaxation it appeared again like a red pin-point. Throbbing between existence and nonexistence, now visible, now invisible, it was the beginning of life.[14]

By the seventeenth century enough people were engaged in science to form societies for the purpose of exchanging information and viewing demonstrations. In England the most important of these was the Royal Society of London, which published the proceedings of these meetings along with letters from other scientists around the world. Some of the experiments were very peculiar, showing an apparent indifference to suffering and other niceties. An investigator reporting on what he believed to be the human allantois said that this belief was supported by the fact that it "is always turgid with a Liquor, that in color, taste and smell, seems Urinous."[15]

Samuel Pepys, who treated his evenings at the Royal Society as wonderful entertainment, on a par with theatre, but with the added pleasure of being intellectually important, reports on a demonstration in which all the blood was drained from a dog and then returned without apparent detriment to the animal. Recall that the first anesthetics were developed almost two centuries later.

Sometimes discretion was required in reporting. Anton van Leeuwenhoek, the early microscopist who discovered the world of infusoria and microbes, also was the first to present a drawing of human spermatozoa. His drawings and description were sent to the Royal Society of London from Holland with a note which, unlike most of the proceedings of the society, was published in Latin rather than English. It said:

> I remember that some three or four years ago I examined seminal fluid at the request of the late Mr. Oldenburg, B.M. and as I felt

averse from making further investigations and still more so from describing them, I did not continue my observations. What I investigate is only what, without sinfully defiling myself, remains as a residue after conjugal coitus. And if your Lordship should consider that these observations may disgust or scandalize the learned, I earnestly beg your Lordship to regard them as private and to publish or destroy them, as your Lordship thinks fit.[16]

This same sense of joyful curiosity and glee at the triumph of clarity in a world full of obfuscations and opacities can be read into Galileo's *Dialogues,* in Darwin, and more generally in the best of scientific writing even today. The "Eureka!" of Archimedes is repeated at any scientific frontier, even minor ones. In my forty professional years I have found a handful of facts and a few bits of valid theory. I don't think that any of it would have not been discovered in due time if I had done nothing, and I am not absolutely sure that others would have found them worth reporting. Nevertheless, embarking on a new investigation has always been a pleasure to me, which is mimicked to a tiny degree by the feeling of walking out on a field of new snow and leaving the first path of footprints. I still feel that it requires a heaviness of head to not want to study the parts of the world that are clearly there, without regard to self-serving opinions. Part of the pleasure resides in the fact that this clarity is just difficult enough to achieve that there is a small burst of pleasure when one succeeds in appreciating a new or unfamiliar fact or theory.

With the passage of time and people the patches of untrodden snow become more and more circumscribed. Even so in the history of science. In the early meetings of the Royal Society of London, the French Academy, or the Italian Academy of the Lynx (named for the sharp intellectual sight of its members), all sorts of facts might share an evening's program. Medical, geological, astronomical, and archaeological facts might be heard in sequence. With the passage of time, each of these areas became so rich in facts and theories that keeping up with new information in geology alone precluded a sophisticated understanding of medicine or astronomy, and the same for the other specialties.

The scientific world began to split. Meetings were subdivided by specialty, and societies were founded to encourage these meetings and publish results in particular fields. Geology was born in the late eighteenth century, chemistry at around the same time, and so on.

My discipline, ecology, only appeared with an identity of its own in the early twentieth century. When I was in graduate school there were around a half-dozen journals in the field, each published monthly or quarterly. By and large I read all of them and had time left over. I had met most of the prominent people working in the field. Now there are many dozens of journals. There are thousands of members in the American Ecological Society and thousands more in the British equivalent.[17] I have only vague hints of the ecological ferment going on in Poland and Spain and know nothing of what may be happening in Yugoslavia. My relation to most of my own specialty has become that of a spectator.

The stream of scientific knowledge is often compared with a swelling river in time, but unlike a river it subdivides as one moves downstream and seems to shatter into multiple rivulets as it empties into the future. The attractive simplicity of science is in constant danger of disappearing.

Specialization tends to conserve scientific simplicity to some extent. At its beginnings, science as a whole abandoned those problems of moral philosophy, esthetics, and religion that are not science. Taking advantage of this simplification, it focused on the empirical, but the empirical world in its entirety is too rich to be understood, so that each scientific specialty abandoned large sections of it to other specialists. This helps until the subsections become overwhelming and further specialization is required. If this were all there were to it, the situation would be highly unsatisfactory; eventually, the clarity and simplicity of science would disappear. Particular facts would tend to be rediscovered under different names in different specialties, unknown to one another. Even a computerized data bank of facts would not help, since searching such a bank requires clear directions and classification.

The exuberant burgeoning and fractionation of factual information bring us back to the complexity of the Tibetan tankas. But just as the disciple passes through the tanka and notes that the thousands of deities have vanished without any loss of important knowledge, so the student of science passes more or less beyond the simply observable facts to a level of theoretical insight that permits the individual fact to vanish, in the sense of becoming completely uninteresting. Science begins with a simplification—stripping away

the nonempirical from our image of the world. But then the empirical world threatens to overwhelm the possibility of thought. Yet
beyond the simple naively discernible empirical fact appears a new
kind of simplifying force which permits us to see the facts as mere
instances of theory. The theories can be even more fascinating than
the facts. Fascination with the intellectual process of theory construction and the theories themselves is again a deeply personal
source of satisfaction and motivation.[18]

The workings of a particular lever are scientifically comprehensible in terms of elementary physical theory and therefore need not
be remembered for any scientific reason. If the lever is made of ivory
or was the one that was used to teach Einstein, it may be of some
interest from other standpoints. Facts that can be subsumed under
theoretical generality lose interest. Facts that deny theory are of
prime interest, and facts that are ambiguous in their relation to
theory have not yet been adequately understood. In principle, theories can organize and simplify the tangle of facts, but we will see
that not all facts take this process kindly.

The role of scientific theory differs among scientific subareas,
which I will call "sciences" for short, with the understanding that
they are all parts of some one big body of science. What seems like
an interesting subdivision need not prove to be so after a bit of
study. Some once respectable sciences have become vacuous. They
flourish for only a little while and then disappear. Who misses
alchemy or phrenology, and who recalls pharmacognosy? Among
sciences that clearly are not vacuous the rate of advance may differ
considerably.

Some sciences focus on what can be understood within a particular theoretical context and stipulate that at least for the time being
they will try to avoid very difficult problems. This has been the
central approach of physics and mathematics, which I will call
"tractable sciences." An alternative approach is to adopt a highly
circumscribed subject matter defined on a commonsense basis and
attempt to deal as completely as possible with that. Geology involves
everything about rocks and minerals, ecology everything about the
interactions of organisms with their environment, and astronomy
everything about stars and planets. They are "intractable sciences."
Particular workers within a science may choose to focus on the

apparently simplest available problems within the science. It is in this spirit that I have focused my research on extremely simple animals.

Tractable sciences focus on a few central problems at any one time. As a rule these problems are defined by their intellectual interest, although the attention paid to particular problems may arise as much from practical pressures or limitations in technology as from intellectual concerns. Solutions of particular focal problems permit the development of new questions.

Mathematics and physics are eminently tractable in part because they have agreed on the ground rule that they can discard certain questions as belonging to the realm of "nonmathematics" and "non-physics" and retrieve them if they should become theoretically clear or fascinating. Sciences defined by their concern with a general subject area are often intractable. For example, medicine, geology, and meterology are intractable. Ecology may be the most intractable legitimate science that has ever developed.

Sciences may become tractable if a sufficiently powerful theoretical advance is made. For more than a century the study of laminar fluid dynamics was part of physics, while turbulent flow was exiled to meteorology or oceanography. With the recent development of new mathematical approaches involving fractals, turbulent flow is smoothly returning to physics. Within biology, biochemistry was a collection of curious facts 100 years ago. In the last forty years biochemistry, genetics, and molecular biology have been clarified and simplified by a series of breakthroughs, related to DNA, proteins, and membranes. There is still active research to be done, but it occurs within focused areas. In modern biochemistry, as in any tractable science, it is usually clear what single important question must be answered next.

I will describe the processes of simplification and minimalization in several sciences. What we will see is an ongoing skirmishing war on the edge of scientific knowledge, in which overwhelming complexity and confusion are just held at bay.

7

Explaining the Whole Universe: Motion, Velocity, and Direction

Hegelian and Marxist dialectic involves a clash of forces creating a synthesis. An alternative image of historic change is that of successive bursts of creative force, breaking the bounds of one creation to produce a new one. The new creation contains vestiges of the old and may contain seeds of a new one by which it will be replaced. The basic notion is easily discernible in the myth of the Olympian gods overthrowing and eating the Titans. It is vividly stated by the sixteenth-century kabbalist Isaac Luria of Safed,[1] known as Ari the Lion. In the Lurianic kabbalah, creation is seen as a process that began with the withdrawal of God from the world, thereby creating space. (If God is endless then space itself must be created!) Then divine forces readvanced into this space to make physical shells containing energy. The energy was too great to be contained in these first physical objects, which consequently broke open and were encased in deeper shells, and for eons the sparks of divine energy have been encased more deeply in shells within shells. The shells hide the internal forces so that one can live in the created world; but at the same time the duty of the saint is to break through the shells to release the divine sparks within them.

The Ari presented a poetic, untestable, emotionally charged vision, but the world does seem to contain actual sparks. When these are released from their shells by intellectual innovators, they set up cascades of new sparks destroying old complexities and creating new ones. Most intellectuals do not do anything that dramatic. Those that play the game of science and in the process invent new games are called geniuses. What had been unknown, or considered trivial, became focal, and what had been considered the central intellectual problems became passé.

Among my intellectual friends and acquaintances are about a dozen officially acknowledged geniuses.[2] Some of them have what appear to me to be deep minds that can clearly visualize images and processes which I can follow only in broadest outline, and then with the hint of a headache. Much of mathematics I find very difficult in this sense, but so do even some very important mathematicians.

It is modish for popularly recognized geniuses to know baseball scores or foreign languages or chess or to be skilled at music, lechery, or something else quite unrelated to their actual work and generally acknowledged to be a proper activity for a "regular guy." Nevertheless, most of the geniuses I have known are distinguished by monomania more than intellectual breadth. They have shown a capacity for holding some one idea, or process, or technique, or problem, at the focus of their effort and attention with an intensity and over a length of time that would be impossible for most of us. In this process the focal problem is refocused. What is extraneous has been defined and then discarded, leaving a simplified world that is then obsessively explored and analyzed. The stroke of genius may be to define the problem and outline a tentative solution, but this is usually only the beginning.

Historians who are so inclined can find a multitude of hints in the writings of earlier, more obscure, workers demonstrating that they understood the central idea that became attached to a later and more famous name. In the tiny village of Bethany, in West Virginia's northern panhandle, a permanently obscure corner of the world, I read on a plaque that the inventor of the telephone, several decades before Alexander Bell, was a man named Dolbear. Italy, Russia, and I suspect many other countries claim other predecessors of Bell; only readers of green bronze plaques will know their names. Actually, a series of lawsuits clouded the primacy of Bell in this invention, but nevertheless the genius of Bell stands.

When Leibnitz complained to the Royal Society of London that his role in the development of the calculus had been slighted in favor of Sir Isaac Newton, the president of the Royal Society graciously established a committee to look into the matter and even agreed to serve as committee chairman. This would have seemed eminently fair had the president not been Sir Isaac Newton, who was using the society as his own political weapon.[3] Charles Darwin seems to have been a much gentler and more pleasant fellow than Newton, but while he shirked the intellectual battlefront, he egged on his friends to fight for him, made sure that all important critics had copies of his papers, cleared his trail carefully before announcing results, and in general behaved in what we would now recognize as a success-oriented mode.

One can find many contemporary examples of keen political acumen among living geniuses, in all fields of intellectual endeavor. Picking one's own referees and critics, choosing an agent, publisher, or gallery, and befriending possible patrons all clearly affect one's attainment of genius status. To indulge just a bit in gossip, one of my favorites is a brilliant woman friend of mine who arranged to write the New York Times book review of a book written by her son from a first marriage.

Nevertheless, intense mental ability, flashes of insight, and intellectual effort have a role. Most of the mere intellectuals I have known assume that the system is ultimately just and do not attempt to manipulate it in the same way as the geniuses. These minor lights tend to write, paint, or do research out of a compulsive need, poignantly hoping to be noticed but doing almost nothing that might achieve notoriety.

Underlying all phenomena are what we call natural laws, and with time these are released and become manifest, changing our sense of the mundane world, but the changed world seems to release even deeper laws and the process continues. In general, "natural laws" are simplifications of nature in the sense that entire classes of observations, with perhaps millions of separate instances, can be subsumed as merely examples of the operation of some natural law. In exchange, new thought patterns may be required in order to construct or understand these laws. Often scientific laws either violate or complicate the concept of "common sense."

The basic thesis of Thomas Kuhn's extremely important book *The Structure of Scientific Revolutions* is that the ongoing history of scientific work occurs in two modes.[4] Periods of "normal science" are interspersed with periods of "revolutionary science." During each period of normal science it is possible to identify a system of acceptable questions, answers, and often procedures, a "paradigm" defining what constitutes proper scientific activity. With the passage of time and the accumulation of effort, paradigms can become un-satisfactory. This may occur for a variety of reasons. The least interesting is that the subject matter has been exhausted, so that there is no longer anything of interest to discover. The classical physics of wheels, levers, and inclined planes was a fascinating and successful area of investigation, and is still exciting to elementary students. Nevertheless, while how to put levers and screws together in an engineering context is still of practical importance, this field is not considered a scientific frontier.

Paradigms may lose their interest if they are ineffective in achiev-ing their original ends. The paradigms of strict Freudian psychiatry dissolved in a welter of case histories and an inability to objectively verify the clinical value of treatments. The ego, id, superego, and even some aspects of the subconscious itself all continued to evade operational definitions independent of the rationalizations of the psychiatrists and their patients. Paradigms may also be killed or overthrown by developing alternative approaches to the same sub-ject matter. Massive concern with pharmacological treatments un-dercut the clinical role of the classical Freudians.

The most interesting way for paradigms to die is at the hands of revolution. When the pattern of questions and answers that are acceptable in a given empirical area have been replaced by a new paradigmatic set, then a Kuhnian scientific revolution has occurred.

Kuhn's book was enormously popular among young scientists of the late sixties and seventies. The image of the intellectual revolu-tionary has always been appealing, and Kuhn's analysis almost pro-vides a blueprint—in fact, a new paradigm—for causing a revolu-tion. A successful revolutionary, a breaker of an old paradigm and creator of a new, is in some sense a more important, and certainly more romantic, person than a "normal" scientist. The recent bitter controversies in paleontology, systematics, ecology, and other of the less-well-formulated areas of biology trace back to Kuhn, as wit-nessed by the sudden appearance and repeated use of the words "paradigm" and "revolution" in scientific papers.[5] It may very well be that vituperative combat may characterize premature revolu-tions—mere revolts—much more than true revolutions. I will return to these topics later.

Tolstoy began *Anna Karenina* with the idea that all happy marriages are happy in the same way but unhappy marriages are each unhappy in their own way. Scientific revolutions also each occur in their own way. There is no real repetition in the cascades of creativity in science. They are richer than the dreams of the Ari. Also, large revolutions set up cascades of minor ones.

Newtonian gravitation, Darwinian evolution, thermodynamics, Einsteinian relativity, and the Watson–Crick double helix model of DNA are enormously different in many ways, but they are arguably the five greatest scientific revolutions of modern science. They shattered the preexisting paradigms not only of science but of common sense. Vast areas became suddenly simplified, at the cost of generating a welter of apparent absurdities that had to then serve as the focus of further work.

Important laws have implications that would not be obvious at all until the laws were stated. One characteristic of successful revolutions is a departure from the simplicity or complexity of the preexisting paradigm. Just as complication becomes an invitation for simplification or minimalism in art and religion, excessive complication or simplification in science cries out for reversal.

The processes of biological evolution, which we will examine in the next chapter, and other irreversible changes in the cosmos produce new problems, whose resolutions require new natural laws as well as revisions of old ones. Not only do natural laws become revealed as a result of scientific work, but even in the absence of human intellectual constructs, natural history—in the literal sense of the sequence of events in nature—actually creates new scientific laws that could not have been revealed by any effort if that effort occurred too early. Geology was meaningless before the Earth's crust had cooled sufficiently to permit solid rock; anthropology had no meaning before the evolution of the primates; and computer science is still being born.[6] This will be clearer after we have examined more science.

For 300 years the Newtonian law of gravity has served as a model for all the rest of science. Its enormous success began the age of science and the myth of scientific infallibility, which is now being rapidly eroded. Gravitation subsumed an amazing array of phenomena. It also radically altered the sense of what makes sense.

The law states that the gravitational attraction between objects is proportional to their mass and inversely proportional to the square of the distance between them. The idea that gravitational pull varies as the inverse of distance squared is not unreasonable. We can visualize two point masses each at the center of a sphere with a radius equal to the distance between them. It seems plausible that the "sense" that each mass has of the other should be related to the surface area of the sphere with the other mass at its center, with its intensity being reduced as the sphere becomes larger, as an image printed on a balloon becomes more dilute as the balloon expands.

The law of gravity does not stand as an isolated equation. It is embedded in a set of physical and mathematical assumptions that are themselves counterintuitive, in the sense of violating previous assumptions. There is not just an isolated law of gravitation, but rather an entire theory of gravity. Some of the implications of the law of gravity were not startling, but some radically altered "common sense," sweeping away concepts of physics and astronomy that had stood for thousands of years.

The idea of a constant acceleration due to terrestrial gravity, independent of mass, is due to Galileo, Newton's older contemporary. The story is that Galileo climbed the leaning tower of Pisa and dropped two cannonballs of radically different weight and thereby experimentally demonstrated that objects fall at the same rate regardless of their mass. This would have denied the common sense of the Aristotelian interpretation of falling—that objects are attracted to the center of the earth, their "natural" point of rest, with a force proportional to their weight. There is no evidence that the ascent of the tower ever occurred. Galileo referred to a *gedankenexperiment,* an imaginary experiment, which he may have felt established the result of independence of falling speeds and weight with such intellectual clarity that climbing towers was an unnecessary effort.

> If two bodies move downward [in natural motion], one more slowly than the other, for example, if one is wood and the other an [inflated] bladder, our assumption is as follows: if they are combined, the combination will fall more slowly than the wood alone, but more swiftly than the bladder alone. For it is clear that the speed of the wood will be retarded by the slowness of the

bladder, and the slowness of the bladder will be accelerated by the speed of the wood; . . . some motion will result intermediate between the slowness of the bladder and the speed of the wood.

On the basis of this assumption, I argue as follows in proving that bodies of the same material but of unequal volume move [in natural motion] with the same speed. Suppose that there are two bodies of the same material, the larger a, and the smaller b, and suppose, if it is possible . . . that a moves in natural motion more swiftly than b. We have then two bodies of which one moves more swiftly. Therefore, according to our assumption, the combination of the two bodies will move more slowly than that part which by itself moved more swiftly than the other. If, then, a and b are combined, the combination will move more slowly than a alone. But the combination of a and b is larger than a alone. therefore, contrary to the assertion of our opponents, the larger body will move more slowly than the smaller. But this would be self-contradictory.[7]

Note that this is a resort to common sense, which nature often violates. Certainly the idea of gravitational force itself violated the common sense of the seventeenth century. If any two objects pull on each other, how is the pull transmitted? Are we willing to accept that two objects interact with each other without any actual contact? Such action at a distance was a strong argument for doubting all of Newton's work, and even made Newton extremely uncomfortable.

Newton assumed the existence of a terrestrial gravitational constant that would transform the proportionalities into equations; then he strengthened the assumption by stating that there is only one gravitational constant for the entire solar system, which he then proceeded to measure.

Since then we have grown accustomed to forces and energy moving across empty space. Radio seems no more surprising than nineteenth-century telegraphy, although one involves a wire and one does not. But even in this blasé age, bright young students in elementary physics classes, after hearing about electromagnetic waves, will ask the rude question; "What is waving?" and the better the instructor the less answer is provided.

My favorite reply was given by Gabriel Weinreich at the University of Michigan, who used the question as an opportunity to tell the story of the two Russians discussing the wonders of science in the early thirties. One asked, "Please, can you explain to me the workings of the telephone?" "Certainly! Think of a long thin Dachshund dog with its head in Minsk and tail in Pinsk. If you pinch the tail in Pinsk the head barks in Minsk." "Thank you and what about radio?" "Just the same but no dog!" Gravity makes no more or less sense than radio but has been with us longer.

Newton stated the law of gravity in terms of mass rather than weight. This is also a major imaginative leap. Mass is not the same as weight. Weight is a measurement on a balance or scale; actual substance is something else. Today, in the space age, this is fairly easy for us to imagine, since we know that an object can be made weightless without in any way diminishing its mass. Weightlessness is a function of an object's trajectory. Weight is an Earth-bound concept. We can ask what an object with the mass of a rocket or a moon would weigh on the surface of the earth, but while they are in orbit these objects must be considered as weightless. But the space program that made that notion into common sense is built on the theory of Newton and not the converse. The intellectual leap to a weightless orbit came from seventeenth-century scientists with feet on the ground.

All bits of matter are attracted toward one another. The strength of the attraction is proportional to the products of their masses and inversely proportional to the square of the distance between them. The trajectories of existing planets are those for which the mutual attraction between them and all the other objects in the universe balance out in just the way which keeps them in reasonably fixed orbits. The relation between any two planets is that if the rest of the universe were absent they would fall toward each other, but in the real situation their other gravitational commitments prevent this. What a wonderful thing to imagine, and what is most pleasing is that it is not a dream or a poetic image. It is as real as apples and moons themselves.

Even conceding Galileo's argument that on Earth there is a constant gravitational force, the universal gravitational constant—a proportionality constant which permitted actual prediction of the interactive pattern between any two objects—was just as implausible as action at a distance. Why should there be a universal number for the entire cosmos, and why should it have some particular value rather than some other?

From Aristotle and before it seemed clear that objects only moved if work was performed on them; if this work stopped, they came to a halt. The theory of gravitation requires that any change in the velocity of an object takes work and that in the absence of such work objects will keep moving at the same rate. At first glance this makes no sense, since if we stop pushing a wagon the wagon stops rolling, but we all learned in kindergarten about friction. We know that a well-oiled wagon on a smooth road will not slow down for a long time. It was Newtonian physics that made friction important in our intellectual curriculum, and not simply a technological problem. A great scientific theory shifts the focus of knowledge, just as much as a breakthrough in art or religion.

From the standpoint of the theory of gravity, what is interesting is change in velocity and direction rather than constant velocity and direction. But to calculate continuous change required a new view of mathematics. Ever since the first account books of the Babylonians, difference between measurements could be calculated. This year had more grain stored than last year. This horse covers more ground in an hour than that runner. What is less obvious is how to indicate that going up a hill the runner slows down and going down he speeds up. A new mathematics, including new notation, for expressing rates of change in velocity itself was needed.

It is still argued whether this branch of mathematics—the calculus of advanced high school students and college freshmen—owes more of its origin to Newton or to Leibnitz. In any case it was needed for calculations of astronomical orbits of the moon and planets, and its amazing success at this made it the focus of centuries of intellectual effort, despite Bishop Berkeley's very plausible objections on the grounds of elementary common sense, made in the eighteenth century. He noted that it is an axiom of arithmetic that any number divided into zero is zero, and any number divided by zero is either infinity or, if you prefer, is pure nonsense. Calculus is concerned with the slope of a tangent to a line. A tangent touches the line at a point, and a point has zero extension. To define the slope as the ratio of the vertical distance the line traverses while it traverses the horizontal width of the point is apparently an exercise in dividing zero by zero. Calculus squirms around this by carefully distinguishing between a number that is zero and one that is approaching arbitrarily close to zero. It required that the result of dividing one number into some other number that was so small that it was approximately zero was not itself zero. To be "approximately" zero meant that for any given actual distance from zero, the approximately zero number was even closer to zero!

Also, calculus only works for systems that are in certain senses smooth. One may calculate the rate of change of the slope of the surface of an orange, but not of a pineapple, since the points on the pineapple's skin represent discontinuities in direction. The pineapple has a skin and a shape, but calculus cannot describe it. These problems made very little practical difference for several centuries. Morris Kline has expressed the opinion that any hope of placing mathematics on a firm, reasonable basis vanished with the development of calculus.[8] As Lewis Carroll pointed out, with practice it was possible to believe ten impossible things before breakfast! We will return to this later.

The successes of seventeenth- and eighteenth-century astronomers at determining orbits and constructing predictive models of the heavens was not much greater than that of the third-century astronomer Ptolemy, who believed the Earth to be at the center of the universe. Ptolemy could get very good predictions with his geocentric model if he assumed the sun and the planets to be moving along several different circular orbits simultaneously. To understand this, think of the path of a speck on a wheel of a car driving in circles on a merry-go-round. This involves three circular orbits. Some seventy-five rotational motions in all were involved in explaining the trajectories of the planets, the speed of each one being estimated from direct observations combined with a set of formulae. By putting the sun at the center, Copernicus could reduce the number of rotational motions needed to around thirty-five; he was very conscious that this simplification was the main strength of his theory, and the reason it was preferable to Ptolemy's.

Newton used one universal gravitational constant and estimates of the masses and relative distances of the objects in the solar system as the starting point, and from these he calculated the planetary orbits. The information going into the system had been enormously simplified, while the calculations had become less intuitive. The magnificence of Newton's accomplishment permitted Alexander Pope to write, "Nature and Nature's Laws lay hid in Night. God said *Let Newton be!* and All was Light."[9] Clocks, tides, comets, planets, and moons all were part of one marvelous pattern. What was left to discover?

What Newton had ignored were those interactions of matter that depend on its kind and organization as well as its mass. The Newtonian revolution had built a minimalist world where the irregular, the nonlinear, the chemical, the living interactions of matter were deliberately ignored. That simplified world then could be completely analyzed, but the simplified Newtonian list of properties of the things in the universe by no means exhausted the complete list. Exploration of these other properties of the universe has occupied science for the last 300 years.

Even astronomical prediction based on Newtonian principles was not absolutely perfect. The ellipses of the planets seemed a touch wobbly. The exceptions to precise astronomical predictions were attributed to human and instrumental errors interfering with proper measurement. Part of the error was also attributed to the fact that for purposes of calculation the theory of gravity usually assumed that the mass of an object could be thought of as concentrated at a point, thereby ignoring shape. (Later, observed deviations of planetary orbits from Newtonian predictions were used to infer the existence of the as yet undiscovered planets Pluto and Uranus. Faith in the theory and the precision of measurement had increased enough to enforce belief in the completely unseen.) In this century the Newtonian model was found to be imperfect, and we now have an Einsteinian relativistic theory of astrophysics, but the enshrinement of science as a way of knowing was precipitated by Newton's theory.

Newtonian astronomy set a standard for the power of scientific theories that has really never been excelled. Physics, mathematics, and astronomy were taken as models for what science ought to be. The simplicity and audacity of their assumptions and the rich diversity of their conclusions were perceived simultaneously as triumphs of human intellect and evidence of the elegance and excellence of nature, and even perhaps as testimonials to the existence of God. Only a combination of instrumental and human measure- ment error, it was felt, and the choice of excessively complex situ- ations for study—and perhaps the intellectual inadequacies of those engaged in the work—prevented understanding of such subjects as chemistry, geology, biology, and human behavior from having the elegance, simplicity, and power of Newtonian physics. Surely, the best path for studying these other more recalcitrant branches of science was to use the Newtonian methods and mathematical tools developed for physics. We will see that after several centuries of not working very well, this viewpoint is presently under challenge.

But stars, and in general the objects studied by classical physics, can be simplified for intellectual convenience without losing their interest for the physicist. Before the mathematical aspects of New- tonian physics could be applied, the real world had to be stripped of all those aspects not essential to the central problem of determin- ing motion, velocity, and direction, thereby discarding most of the subject matter of science. A philosophy professor at Yale in the late 1940s objected to this approach. Physics speaks of, perhaps, a cow sliding down a hill. It does not ask after the well-being of the cow— is it a happy and healthy cow? It even need not realize it's a cow at all—it is merely a mass. The mass need not even have the shape of a cow. In fact, before they consider it sliding down the hill the cow is simplified to a "point-mass" and then if the hill is sufficiently smooth and a few other things are all right, we proudly announce the acceleration while sliding down this hill. But is a point mass sliding down an inclined plane the same as a cow sliding down a hill?

The whole argument struck me as silly when I first encountered it, and it does miss the point. Newton had brilliantly left out all that was unessential to motion and direction. But there is a nonsilly aspect to this objection as well. On one hand, the development of mathematical theories often hinges on great simplifications. We can agree with Newton that the acceleration of a cow on a slide depends on the gravitational constant, coefficients of friction, the mass of the cow—but not on all the other things that make the cow inter- esting. On the other hand, the cowness of the cow also requires explanation as much or more than its slide. This will be the subject of our next chapter.

8

Explaining the Rest of the Universe:
Darwinian Insight

In previous chapters I have not presented detailed discussion of particular scientific theories, in part because physical and chemical sciences require an explicit foundation in mathematics, or at least a general level of comfort with quantitative notation. In this chapter we look more closely at the curiously commonsensical evolutionary laws that purport to "explain" all of the 30 million more or less different living "species," but which do not require mathematics beyond multiplication and division, both of which are even named as cognates of farmyard biology.

Defining species and classifying them is a subject in itself. To discuss its details would be an excessive digression. In the following discussion, species of sexually reproducing organisms are defined in the most generally accepted way as groups of organisms that are reproductively isolated from all other groups. How to define asexual species is more difficult and for our purposes not really necessary. The total number of species on earth is somewhere between 10 million and 100 million. (I used 30 million in the paragraph above.) Each species consists of from a few dozen to several billion individual living organisms, each of which differs from all the others. Interspecific differences may be as extreme as that between moss and moose, as slight as that between humans and chimpanzees, or as subtle as that between many of the species of Hawaiian fruitflies.[1]

I have suggested several times in this book that scientific explanations for phenomena generally appear simpler than the phenomena themselves. This is part of the intellectual appeal and general utility of science. The complexity of a factory pales besides that of essentially any organism, and the physical laws used in building the factory are in some ways considerably more complex than the laws of biological evolution which, in one sense, underlie the organism. Discussion of evolutionary theory is pivotal to our theme, since it may be the prime example of an extremely simple natural law that can explain highly complex phenomena.

DNA STRUCTURE

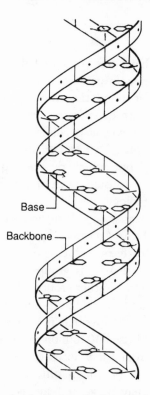

Figure 5. The DNA molecule is a long double-stranded chain. Each strand is made up of millions of minute subunits called nucleotides. A nucleotide is composed of three parts: a sugar, a phosphate group, and a flattened structure called a base. The sugar and phosphate group of each nucleotide contribute to the backbone of the DNA strand. The backbones of the two strands wind around each other to form a double helix. The bases, which are perpendicular to the sugars, tend to stack one on top of one another, much like steps in a spiral staircase. The four bases in DNA are adenine (A), guanine (G), thymine (T), and cytosine (C). The information carried by DNA is coded in sequences of nucleotide bases.

Also, the simple idea of evolution has generated tidal waves of misunderstanding and controversy for more than a century, casting doubt on the generality and utility of common sense and forcing more careful examination of the concept of simplicity. In fact, with the theory of evolution the gamelike character of simplicity and complexity (as described in Chapter 3) breaks down. People who discuss evolution seem to have a very difficult time keeping to a single playing field. In that sense, consideration of evolution provides an introduction to the emotional and political loading of the idea of simplicity, which we will discuss in Chapter 9. Also, the discussion of biological evolution will provide raw material for the philosophical analysis of extreme complexity, how it may originate, and how it may be understood (Chapter 10).

The Complexity of Organisms

Merely visiting a zoo or a garden, or doing gross dissection of the sort done by butchers, cooks, and elementary students of biology, makes it apparent that organisms are complex compared with inanimate objects. In fact, to assert that organisms are the most complex objects in the known universe is not hyperbole. Organisms also seem to differ in anatomical complexity in obvious ways. The plumage of a peacock or Argus pheasant seems more complex than that of a crow, the flower of an orchid than that of broccoli. Compare on one hand the polyps and jellyfish, which have no true limbs, head, or even anus, with vertebrates, segmented worms, or insects. The polyps seem amazingly simple, but they have been successful in dealing with their problems for several hundred million years without the need for more complex anatomy. In some sense the polyps have been more successful than those that have been forced to abandon old ways of living and to experiment with new ones.[2]

The complexity of organisms certainly persists at all magnifications so that, while early astronomers were enlarging the subject matter of science to the unimaginably large, biologists were becoming fascinated by the study of the unimaginably small. Swammerdam, Leeuwenhoek, Hooke, and others, using the relatively crude microscopes of the seventeenth and early eighteenth centuries, made beautiful anatomical studies of caterpillars, worms, fleas, and even of "infusoria," organisms too small to be seen with the naked eye. These tiny animals appeared to have anatomy as complex as larger ones.[3]

In fact, revelations of surprising structure in the very small parts of organisms continue to this day. In their individual cells, their most internal parts, crows, mosses, jellyfish, and infusoria are very similar to peacocks, orchids, or statesmen. Gross outer anatomy, however important in dealing with their particular life's problems, is built over a deeper, and, as we will demonstrate, older complexity, which has only been explored in recent decades.

When I was a student we were given the impression that the cell contains a kind of amorphous jelly in which float a few other structures like vacuoles and the mysterious Golgi apparatus. In the middle of each cell was the nucleus, containing the chromosomes on which rode the genes. (Perhaps astride?) Further investigations with better microscopes and preparation techniques have replaced this simplistic picture. Now we know that any single cell, when examined at high enough magnification, shows fantastic details of structure and compactness of design that make the finest mechanical devices seem crude.

Most cells of animals and plants may be thought of as miniature factories, each containing many little workshops enclosed in a maze of membranes. Each little shop processes, combines, shapes, or sub-divides molecules and then passes the products through to the other side of a membrane, where they are further processed. This permits sorting of materials and prevents a backlog of reaction products. At the borders of each cell, material is entering and leaving, as if from the loading docks of a factory.

The directions for all of this activity are in the cell's library of genes, the famous DNA molecule. Molecular messengers, under appropriate circumstances, read out from the sequence of bases in the DNA a set of instructions for chemical processes. Amazingly enough, they only read from those genes whose information is needed by that cell at that point, the rest of the genes lying about like uncalled for but hopeful books in a library. The DNA of even a small bacterial cell contains much more information than this

book, arranged in a wonderfully compact linear form. Some sequences of bacterial DNA can be read in two directions and provide
meaningful but different messages in each direction, while other
sequences contain multiple messages, depending on where one starts
reading. Most cells also even contain apparatus for the upkeep and
repair of the genes themselves.

But before we put too much weight on the image of cells as mere
factories, recall that there are many types of cells, each with a
special kind of function. Some cells become linings to surfaces such
as external skin, the surface of internal organs, the internal linings
of digestive tubes and lungs and gills. Many types of these epithelial
cells either secrete chemicals into tubes, or onto surfaces, like those
that pour out digestive juices into gut linings or cooling sweat onto
hot skin. Some cells may become detached from the surfaces to act
as lubricants, as in our mouths, or as tiny protective scales, like our
outer skin. Other types of cells become motile, either becoming
detached and moving like amoebae through the body, as some of
our white blood cells do, or retaining attachments while pulling on
other cells, as muscle cells do. Some kinds of muscle cells initiate
their own contractions, others wait for nerves to signal them; in
some organisms one end of a muscle cell can act as its own nerve
and the other end can contract. Some cells absorb light energy and
store it in chemical form, others secrete gases, and on and on.

Despite their differences, all the body cells in a particular organism contain the same genetic information (except for occasional
misprints called "mutations," which are themselves extremely important). This genetic identity implies that the development process
which gives rise to differentiation of different cell types and to the
organization of combinations of these into organs must involve a
very selective reading from the available genetic information. These
developmental processes are the domain of developmental biology.

All free-living organisms, single-celled or multicelled, perform
essentially the same processes in staying alive. They feed, respire,
incorporate some molecules into their bodies more or less unchanged,
split and rearrange others, synthesize new molecules from old, and
eliminate waste products. In multicellular organisms the different
activities required to be alive can, to some degree, be performed by

TRANSLATING DNA INTO PROTEIN

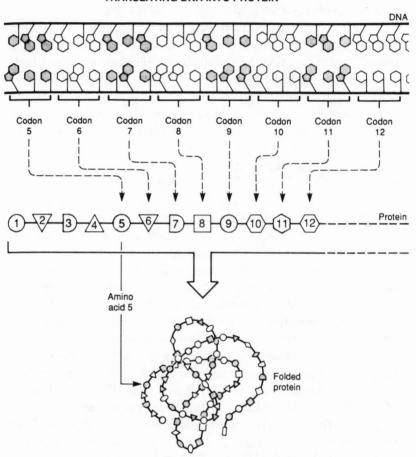

Figure 6. Information in DNA is stored in a series of three-letter "words" called codons (the "letters" are individual nucleotides in the DNA molecule). Each codon specifies a particular amino acid. For example, codon 5 has information for the fifth amino acid in the protein. The overall shape and activity of a protein depend on the precise order of amino acids. The information for the order is stored in the DNA.

different types of cells, each contributing to the overall performance of the entire organism. Large mammals, such as people and dogs, contain on the order of a billion cells, each of which interacts with others.

In organisms that are not subdivided into cells, for example the ubiquitous paramecium, different parts of the same cell have different functions. In a single paramecium are contractile threads which serve the purposes of muscle cells, and a system of tiny "vacuoles" (membrane-enclosed vesicles) and canals that serve as kidneys, and other vacuoles that have a gutlike function. These subcellular structures are called "organelles." Protozoan cells are generally rich in such organelles.

From one standpoint, the complexity of those cells that can function as free-living, independent organisms is considerably greater than that of each of the cells that constitute parts of multicellular organisms. In exchange, multicellular organisms have organized their cells into functional and anatomical parts so that the aggregate is organized to perform all the functions of life. How meaningful is it, then, to grade organisms as to their complexity or simplicity?

We will see in Chapter 10 that the "meaningfulness" of statements about complexity and simplicity is particularly difficult to understand if we focus on objects themselves rather than comparisons among objects, and also that all such comparisons require careful ground rules and local definition of what "simplicity" means in the particular context. Nevertheless, there is a common usage among biologists which ranks organisms in complexity, although the ranking of organisms or even of organs on the anatomical level usually does not take cognizance of cellular complexity, and the ranking of complexity on the cellular level does not take account of organs or whole organisms.

Bonner notes that both the number of cells and the number of types of cells are generally correlated with body size.[4] He has suggested that the number of types of cells is a measure of biological complexity, and that larger body size permits a greater number of cell types, which in turn permits more and more kinds of cellular interactions. Some cells pour out secretions which change the chemical composition of the body fluids, thereby altering how other cells function. In the case of neurons, they may have tendrils which reach out and touch, or almost touch, other cells and transmit specific chemical or electrical messages. Even with the smallest of transistors, and the most miniature circuitry, a computer with as many possible interactions among its components as are found in the brain of a pigeon would be larger than the largest supercomputers now available, perhaps the sum of all of them.

The Simplicity of the Theory of Evolution

While wonderful structures and processes are coordinated in any living organism, all organisms are functionally imperfect. All eventually die for one reason or another—accidents, disease, predators, or just the wearing out of parts. The final biological test of how well all this information and structure have been put together to meet the problems of the organism's environment is whether or not that organism will be an ancestor, that is, whether or not it has succeeded in passing on its genetic information to new organisms that will be its replacements.

Organisms so well adapted as to live forever would not need to reproduce, if success is to be measured merely by the persistence of genetic material over time. In fact, that is a very good criterion for measuring evolutionary success. Attempts to assert any characteristic of evolutionary success other than persistence of genetic material through time lead to nonsensical conclusions. Among most organisms, we can see an inverse correlation between life span and individual reproductive rate, as if maintaining genes in one's own body and passing them on to offspring were somehow incompatible.[5]

For example, consider the risks incurred by a bird feeding its young, or the physiological drain caused by lactation. Using persistence of ancestral lines as a definition of biological success, we might conclude that all of us—each bacterium, beetle, elephant, or person—are the descendents of a line of successful ancestors that began at least 3 billion years ago, before all but the oldest extant rocks were formed, before the continents were recognizable, and before the sea was fully salty.[6]

From these ancestors all living organisms carry an ancient legacy.

Unlike those who claim romantic descent from medieval kings, organisms have real proof of their ancient lineage, in the form of bits of genetic information carried in their bodies which they inherited from these early days. It is for this reason that many biochemical properties are common to all life and why isolated bits of genetic material from mammals can function even in bacteria. A fine example of common ancestry is a sequence of 306 bases on the gene called histone 4 which, except for two bases, is absolutely identical in cows and peas. The closest common ancestor of cows and peas lived approximately 2 billion years in the past.[7]

Stop for a moment and consider what wild statements I have just made. I asserted my belief that life spontaneously appeared on Earth billions of years ago and that a common ancestor of cows and peas once existed! Note that I am not convinced that life arose only once, but I am reasonably certain that all present organisms have only one common ancestor, and I'm quite sure that peas and cows shared an ancestor. As a modern biologist, I wrote these apparently absurd assertions with as strong a sense of certainty as if I had written, "The sun will rise tomorrow." In fact, the idea that all organisms are related in a gigantic familial assemblage is so central to modern biological thinking that I cannot really consider the relative complexity and simplicity of organisms, or most of their other properties, without assuming that evolution occurred. I will now present the bases for my certainty.

The basic theory of evolution through natural selection burst on the world in 1859 in Charles Darwin's classic bestseller, *On the Origin of Species*. Darwin's name is universally and properly attached to the theory. Of course he had scientific predecessors. For example, Alfred Russell Wallace described the fundamental evolutionary mechanisms in a letter to Darwin before Darwin had published on evolution, though Darwin's notebooks and correspondence show that he had been working through the details of the theory for decades. Wallace's letter expressed the theory more clearly than Darwin's first publications did, and on July 1, 1858, the ideas of both men were presented at a meeting of the Linnean Society of London—a form of simultaneous publication. Other scientists had grasped the fact that evolution had occurred and had understood one or more components of the mechanism of its occurrence well before Darwin; these included Erasmus Darwin (Charles' grand-father), Lamarck, and others. Speculations on the possibility of evolution go back to the beginnings of literacy. Writing about the history of, and influences on, evolutionary theory, both before and after 1859, is a major scholarly industry, but it is not our subject.[8] For our purpose we can get by with only the elements of the theory very much as it appeared in 1859.

(Evolution is so much a part of our intellectual culture that it is very difficult to consider it without preconceptions. It is probably important to pause and to attempt to reconstruct a kind of inno-cence, perhaps with a cup of coffee, before reading further.)

The theory of evolution, essentially as described by Wallace and Darwin, and as still taught in modern biology courses, is as follows:

(1) Every living organism is the result of a reproductive process, so that all organisms have ancestors, but not all organisms necessar-ily reproduce so as to become ancestors. Also, some reproductive organisms are more successful at being ancestors than others, in the sense that they leave more descendants than others. (Producing more "descendants" is very different from producing immediate progeny. A tree that produces many seedlings, most of which die of blight, is a less successful ancestor than a tree that produces many fewer seedlings which are blight-resistant and which survive to produce seedlings of their own.)

(2) Even very closely similar organisms may differ in their relative success at being ancestors. Success at being an ancestor depends on very slight differences of heredity, development, and environmental circumstances. Certain organisms are said to be *selected* as ancestors, on the basis of their physiological and behavioral capacities and on their good luck in encountering circumstances with which they could cope reasonably well.

(3) To the degree that relative success at becoming ancestors is correlated with genetic properties, there occurs continuously on Earth a *natural selection* in the sense used in biology textbooks. That is, at any point in time the hereditary properties of a population are more similar to the hereditary properties of its more successful rather than its less successful ancestors. The idea of "hereditary properties of a population" must be understood in a technical sense relating to what genes are present, how they are distributed among

organisms, how they are combined into strands within the cells of individual organisms, and perhaps how mating systems operate. Full exposition of these points would fill another book. In fact, it already fills many. Full exposition is not needed for our purpose. In a pinch we could even get away with the elementary assertion that organisms often tend to resemble their parents and closer relatives rather more than remoter relatives.

(4) Parents transmit genetic information to their offspring as coded molecules of genetic material. Occasionally genetic material changes or new material is introduced. This novel genetic material, called a *mutation,* may be transmitted as well, along with preexisting genetic material, to descendants.[9] The differential success of population members at being ancestors of organisms that carry some of their genetic material is called *fitness.* Different phenotypes (that is, bodies, as opposed to genotypes, which are essentially DNA) have different fitness. Since fitness is determined, at least in part, by the genes one inherits from one's ancestors, descendants tend to resemble the fitter members of a population. As we will see below, this information was not available to Darwin and Wallace.[10]

Comments on the Implications of the Theory

I have now presented the basic theory of Darwin and Wallace, with a bit more of modern genetics added. These four brief assertions seem trivial and are generally applicable in a dull way to all organisms. But since organisms differ so greatly, it is not surprising in retrospect that any suggested mechanism for evolving the full array of life must be very general indeed. If this were not so, the mechanism would not apply to all organisms. In fact, properties that permit evolution are among those by which we distinguish between life and not-life. What seems less obvious is how, or in what sense, these four assertions can in any way explain all the complexity and diversity of more than 30 million species. The answer to this question is really only possible on a case by case basis, which is too much for us here. Some generalizations are possible, but even a full selection of these would require another, thicker, book. All I will do is present a few major subareas of the general answer, choosing some which arbitrate in a general way between the simplicity of the initial theory and the complexity of its consequences.

The first two postulates listed above are fairly obvious and absolutely general. The third introduces the idea of a "population," which is a central concept in the theory of the evolution of sexual organisms. A population is a collection of organisms which can, in principle, share common descendants. Evolutionary change on the very small scale involves statistical differences among populations in different places or in the same population examined at different times.

To avoid misconceptions I must focus briefly on the sources of variation among organisms within a single population. I am the father of identical twins. On rare occasions, when they were very young and had deliberately switched clothing, I have confused one for the other. Usually I have had no trouble in telling apart these genetically identical individuals, who were raised under very similar conditions. One was born a bit larger, and after 30-some years is still slightly larger. They differ in mannerisms, politics, attitudes.

As my twin sons originated by the splitting of one fertilized egg, so an armadillo's eggs may divide in four, and the mother may give birth to identical quadruplets. The aspens in a single stand are often genetically identical, as are the plants in a clump of strawberries or violets. In all these cases, there has been some kind of subdivision into genetically identical fragments which each became an organism in its own right. As a rule, the resulting organisms are nevertheless visibly different from each other. This demonstrates that even if there are no genetic differences at all, and even if environmental conditions have been very similar, organisms may differ one from the other as the result of minor developmental differences. Organisms that are not genetically identical will usually show much greater phenotypic differences.

Usually individuals in the same population of organisms are genetically different. The preponderance of these differences arise from a reshuffling of preexisting genetic material, in somewhat the same sense that card hands differ because they come from a reshuffled deck.[11] There is, therefore, an enormous pool of genetic differences among individuals due to recombination of genetic material. The predominant effect of natural selection in any generation is to alter the relative frequencies of preexisting genes, but only to the degree that genetic differences among individuals are not blurred by their developmental difference and by their exposure to differences in the environment.

While the first three postulates would permit change in the properties of a generational sequence of organisms in response to some environmental changes, the change could not exceed the limits set by the initially extant genetic material. This recombination of ancestral genes is necessary but not sufficient to explain the origin of all species from protozoa to penguin. Therefore, without really having any certainty of how it might work, Darwin and Wallace both postulated that new hereditary material occasionally appears by a disruption of some kind in the old material. Unfortunately, Darwin, in later editions of the *Origin,* suggested that new genetic material came from a system of "pangenes," a shot in the dark that did not hit anything. But what really mattered was that Darwin saw that some source of new variation was needed. If really new genetic material can arise, then there is at least the possibility that drastically different-looking organisms might have had a common ancestor. The fact that the mechanism was not clear to him, and that he proceeded to theorize nevertheless, is to Darwin's credit. Almost a century was to pass before the mechanism of the origin of mutations was at all understood.

A mutation is a change in a gene. It is relatively sudden and may occur in any cell of any organism. Mutations are basically random changes in the genetic directions that guide the processes occurring in cells. The *rate* of mutation can be predictably increased by radiation or chemicals or sometimes other genes that interfere with the gene replication process, but usually the direction of mutation is not predictable. And certainly the direction of mutation has nothing to do with the immediate needs of the organism.[12] Mutations are best thought of as errors or misprints in gene duplication.

By and large I contain as many copies of the genes I received from my parents as I have cells in my body. A mutation may occur at any moment in any one of these thousands of genes in any one of my millions of cells. Most of these mutations will make no physiological difference at all. Perhaps they may do some local damage—a mutation in a skin cell may produce a skin cancer—but by and large the mutations that occur in my body cells make almost no difference at all from an evolutionary standpoint; that is, they don't affect my capacity to be an ancestor. The major exception is mutation that occurs in cells which may grow into sperm, fertilize an egg, and become my contribution to the single cell that will develop into a child—because the genes in that single cell will be replicated in every cell of the child's body. The genes in the one sperm that fertilizes the egg—the one chosen by chance out of the many millions—is my share of the book of directions for the new offspring. The genes in the cells of my ear or liver are copies of it; they may be of interest to the ear or to the patch of liver cells, but any mutational errors that occur in these copies will not generally concern the rest of my body and certainly are no concern to my present or future offspring.

Random changes in a carefully edited text or set of blueprints may actually be an improvement, but the chances are that they will not be. Yet it is also true that most misprints do not really matter very much, unless they include rather large portions of a text or critical single letters.[13] Nevertheless, copying errors are the ultimate source of the new genetic material that makes evolution possible.

This sounds most unlikely until we recall the enormous lengths of time during which useful mutations might have arisen, the enormous number of organisms that may have been born during that time, and the fact that all dangerously incorrect genetic misprints leave no descendants. It may help to think of evolution metaphorically as a kind of learning by a population's genetic system, using the bodies of the organisms as a way of testing the environment. Consider a nineteenth-century anti-Darwinian verse:

> There once was a clever baboon,
> who always blew down a bassoon,
> He said: "It appears that
> in millions of years I'll
> eventually hit on a tune."

And in fact if the poor baboon was reprimanded strongly for less tuneful notes and rewarded for the better notes and note sequences, note by note, and if he were then made to repeat the process, starting with the corrected version of his previous attempt, he would hit on a tune fairly quickly. And of course on the scale of the evolutionary process there are always replacement baboons if this one doesn't shape up. (If this kind of argument doesn't help you, skip it and read on or check with a standard text.)[14]

Summarizing the role of hereditary material in evolution with brutal brevity, we might say that organisms of the same population can be shown to differ in their genetic material, even if these differences are not expressed in visible differences of anatomy, phys-iology, or behavior. In each generation, a small fraction of these genetic differences may be due to new mutations, but most were inherited. In each generation, organisms are born whose combina-tion of genes differ from their parents, siblings, and other relatives. Sometimes environmental circumstances are such that some carriers of particular genes or combinations of genes are more successful at becoming ancestors than others, and in that sense are selected.

It is extremely important to emphasize that there is no reason to conclude, simply on the basis of the fact that two organisms differ in anatomy, physiology, or behavior, that the differences between them are of genetic origin. Failure to keep this in mind has been responsible for tragic errors in the application of evolutionary theory to human affairs.[15] Each organism's set of genes may be thought of as directions for the growth and function of the organism itself, with the immediate and important stipulation that the directions need not all be very explicit. My genes directed my development to produce blue eyes, but they left to the developmental processes and environmental circumstances whether or not I joined the Demo-cratic Party and spoke English.

Most mutations are not particularly useful when they first appear in the genome, nor can they persist if they are particularly harmful because they would be eliminated by selection. But sometimes mu-tations being carried as fairly rare genes in the general welter of genetic variation turn out to matter if the environment changes in odd ways. When that happens, evolution does not have to wait for the occurrence of new, adaptive mutations. For example, when the insecticide DDT first came into general use during the Second World War, it was sprayed on natural fly populations, which rapidly became resistant. It is believed that this resistance occurred so rapidly because, even before DDT, out of every several million flies one or two carried particular mutant genes or a rare combination of genes which made them at least somewhat resistant to DDT, a chemical that was not developed until the twentieth century! It was some of these few rare flies that became the ancestors of a large share of the fly populations after the DDT sprayings. Also, the development of resistance to antibiotics among bacterial populations does not apparently wait on the occurrence of new single mutations but rather is due to the preexistence of sufficiently genetically diverse bacteria that out of every 10 million or even 100 million bacteria, one or two are already resistant to the antibiotic. However, most early antibiotics were compounds extracted from molds and had certainly occurred naturally in the past.

Applying DDT and antibiotics are severe and unnatural environmental changes with dramatic effects on their targets. They were intended to destroy their victims. Nature does not have intentions, certainly not the basically malicious intentions of insecticide and bactericide. The overwhelming preponderance of natural selective forces are considered to be much milder—a slightly more severe winter, a change in a drainage pattern, an early or late appearance of a food plant—but the principle of selection remains.[16] For any given set of circumstances that occur, some individuals already have some genetic propensity to take advantage of them more than others or to be damaged by them less than others. These individuals have a much better chance of becoming the ancestors of more than their share of the next generation, and their genes will be more than their share of the population's genes.

There are organisms, times, and places in which evolution may be slower or faster than average. If massive geological or climatological or ecological changes occur, the rate of evolution will be accelerated, as will the rate of extinction of species. If these properties are constant, evolution will be slower. Notice also that while evolution does not have to wait for new mutations following an environmental change, the fact that new mutations occur continuously ensures that the process of evolution will never completely cease because of an exhaustion of genetic variation.

That is the basic theory of evolution. I will add a little more detail about specific aspects of it, but only for the sake of providing an overview, not a proper review, of what developed after Darwin.

The Origin of Species

So far I have said nothing about "the origin of species," and oddly enough Darwin said very little about it in the book of that name. A detailed explanation of the current state of thinking about speciation would be too lengthy, but the subject cannot be omitted completely. For our purpose I will accept the standard view.[17] Genetic diversity among a set of populations of organisms that have descended from a single freely interbreeding population may lead to differences that prevent, or at least hinder, one or more populations from breeding with the others. This may happen because the individuals of the different populations adopt different breeding times or courtship behaviors, or develop different gamete surface chemistries or developmental mechanisms, or for any other reason that makes it either less likely or less advantageous for them to breed with members of some population other than their own. These differences among populations most commonly arise concomitantly with geographic isolation of the populations, although the actual absence of interbreeding can only be demonstrated in nature if the several populations have descendants that have reestablished spatial propinquity and yet do not interbreed.

The mechanisms that set up barriers to interbreeding between incipient species—that is, populations that are on their way to becoming separate species—are not generally of great life and death importance in themselves. The separation of species is a consequence of selection for other things. At the same time, the branching of populations of organisms into separate species permits a somewhat finer tuning of the genetics of particular populations to environmental circumstances.[18] If herring could interbreed with flounder, we would anticipate problems for their offspring.

Despite my informal tone, I am absolutely certain that these are the basic processes that make evolution occur and that the present diversity of eukaryotic organisms is the result of these processes going on for around 2 billion years.[19]

Time and Evolution

Most scientific laws are concerned with short-term processes—a falling weight, a swinging pendulum, a burning match, or even a cow sliding down a hill. Geology and the theory of evolution are concerned with explaining states of nature that may have required enormous lengths of time to develop. A craggy landscape is the product of the long-term operation of geological processes, but the time for developing a mountain range is minuscule compared with that of evolution. The diversity of organisms on Earth is an extreme example of the long-term effect of small events.

This does not mean that all evolutionary change is so slow as to be invisible. The evolution of resistance to DDT and antibiotics has occurred in recent decades, and by some criteria the rapid changes of plant and animal crops in the hands of agricultural experts might be considered a kind of evolution (through *artificial* selection, to be sure). Lest it be considered that these examples are too unnatural, we have more natural ones available. (Notice that any example I cite of evolution during the last half million years is subject to the objection that since humans have been on the planet for that period of time, the world cannot be considered natural. I will simply ignore such carping complaints!)

Most of the little brown birds of American cities' streets are English sparrows, so named for the very good reason that their ancestors were all imported from England. Like those of so many other Americans, their ancestors were released onto the streets of New York City around 150 years ago. From there they have spread over much of North America. In that brief time clear color differences have evolved among English sparrows in different regions of the United States. The birds from the moist forests of Oregon are a darker brown than those in New York, and birds from the pale desert soils of the dry southwest are definitely lighter.[20] In every case the birds have developed a slight color shift consistent with the hypothesis that they are colored to match their environment.

The differences are not dramatic, and no new species has formed, but they are just what we would expect from evolutionary theory, and they have arisen in the last 150 years. There are other similar examples. But notice that the humans who immigrated to America and were set free in New York City at the same time have not shown any evolutionary changes at all; this implies that the timing of evolutionary events must be dealt with case by case, species by species.

How long does speciation take? Bear in mind that speciation is not an inevitable process; some species may not have branched into new species for millions of years. We are understandably curious about how rapidly species can be produced. One remarkable example that seems to give us some sense of time is the occurrence on the Hawaiian archipelago of approximately 300 species of fruitflies, none of which occurs anywhere else in the world. The details of the investigation of these animals are absolutely fascinating but too technical and too lengthy to include here. Summarizing briefly, the arrangement of chromosomes of these animals makes it clear that these species evolved from one or two species of original invaders of the islands. The islands have been in existence for less than 5 million years, and some islands are much younger. A time for speciation of the order of 1,000 to 10,000 years is certainly consistent with the data.[21]

However, this example presents a difficulty. While no single island is much older than around 5 million years, the archipelago is perhaps two or three times that old. Old islands, possible sources for the present fly species, have sunk beneath the sea. We may therefore be seeing the result of speciation events that are older than the present islands themselves. This case illustrates how much care is needed in interpreting evolutionary information.

The multitude of cichlid fishes that occurs in the great African Lakes is a better example. In order for the number of species now found in Lake Malawi to be produced, the time for at least some speciation events must be less than 5,000 years, and maybe as short as a few centuries.[22]

Evolution has not stopped. In fact, since selective forces are strongest when the environment is changing most rapidly, both extinction and speciation are now occurring at a more rapid rate than at almost any other time in the history of life. Extinction of a population is a much more likely effect of environmental change than speciation, just as deleterious mutations are much more likely than beneficial ones. Sometimes we know the moment of extinction. Martha, the last passenger pigeon, died in the Cleveland Zoo in 1914.

The possibility of extinction of at least some species is as well attended as the potentially deadly illnesses of kings and dictators. The whooping crane, black footed ferret, tiger, and giant panda are all so close to extinction that experts have gathered around them and are trying various drastic remedies. Often these efforts are symbolic rather than practical. When the bamboo died in China and the pandas were starving, then-president Reagan personally brought in his luggage several stalks of bamboo and presented them to the People's Republic of China. No one even smiled.

Extinction of species is not inevitable. Despite various contrary conjectures, we have no evidence whatsoever that there is a clock ticking away the life of a species as there is a clock ticking away my life and yours. Extinction is due to circumstances, and one hopes that these fine animals pull through their crisis. But biologists mourn as strongly for the lonely extinctions of snails in the rivers of the Appalachians and insects and plants that have never even been described but are becoming extinct as we destroy the rainforests of Brazil.

I would be pleased, but not startled, if it could be shown that some particular species originated last year. The origin of a species is a less conspicuous and more subtle and less time-determinate event than extinction. Every intermediate stage on the path to possible speciation is known. In some species of animals or plants, different populations show slight statistical tendencies to not interbreed. Other conspecific populations show more complete breeding bar-riers. Once breeding barriers are complete, never again will the two populations share common descendants. They are now separate species. Each of them, like the partners after a divorce, is free to move independently through time, but they always carry vestiges of the time they were united. It is not possible, as suggested by a nineteenth-century skeptic, to sit and watch an onion turn into a lily. However, we can be quite certain that onions and lilies came from a common ancestral population which was not ancestral to carrots or barley or oaks.

Mathematical Evolutionary Theory

The application of evolutionary theory to actual situations, taking cognizance of the full range of biological and ecological peculiarities, is perhaps possible but terribly difficult for most cases, because of the number and variety of the anatomical parts, the richness of the behavioral repertoire, and the long list of environmental properties that make a difference.

If one's vision of the world can be sufficiently simplified the fascinating possibility is opened of reconstructing evolutionary theory in rigorous mathematical form. Starting in the first half of the twentieth century highly formal and difficult mathematical models were constructed for certain subsets of evolutionary phenomena. Among the most brilliant of their creators are Sir Ronald Fisher, J.B.S. Haldane, and Sewall Wright.[23] Their models, and those of their many successors, were primarily concerned with predicting gene frequencies of populations rather than the properties of organisms. Anatomical, physiological, or behavioral properties of organisms were used only as identifying tags or as ways of defining particular properties for genes or complexes of genes. The problems of population genetics are usually simplified by establishing *ceteris paribus* assumptions about all that is not part of the formalism of the theory. Occasionally, by choosing simple organisms, it is possible to develop mathematical evolutionary theory without having to ignore the properties of the organisms themselves.[24] In these cases, however, the range of organisms for which the mathematical theory holds in any rigorous way is very limited.

Even equations and explanatory graphs seem able to break loose from their origins and develop a life of their own. For example, Provine has demonstrated that the Sewall Wright Adaptive Landscape, a classical visualization of genetic change in evolution as movement on a mathematically defined mountainous surface, changed meaning even in the mind of its creator, and certainly in those of his followers who were attempting to prepare canonical texts.[25]

Is Evolutionary Theory Really as Simple as it Seems?

The disparity in complexity between the ongoing laws of evolution and the products of the action of these laws is a wonder and a delight. If the ability to dispel complex confusion by simple under-standing is a measure of scientific elegance, then the theory of evolution is of the greatest elegance that I can imagine. It is a scientific theory whose validity is quite certain. However, unlike anatomy, genetics, embryology, biochemistry, or physiology, all of which can be thought of as components of the story of evolution, the theory of evolution itself acts as a lightning rod for controversy. Looking briefly at some of the controversy will shed light on our main theme. Why the controversy about evolution, more than about its component parts? How can all the pieces be noncontroversial but the final theory be so disturbing? The problem relates to religious fundamentalism and the centrality of humanity in the pattern of nature.

Dozens of quite different stories of how the universe was created have come down to different peoples from dim antiquity. For ex-ample, we know the Hebrew Bible, the Gilgamesh epic, the Enlil story, Greek mythology, the Polynesian legend of Maui. Any com-pendium of comparative anthropology from the recent books by Campbell back to the *Golden Bough* of Frazer and even the historical writings of Herodotus will provide many more.[26] They all share the general theme that some supernatural force or person had a partic-ular interest in Earth and in the people on it and that the rest of the universe was made for the convenience or adornment of Earth and its people.

The idea that the created world is a witness to the intimate relation between people and their creator can generate fine poetry, but like all poetry it may not translate well.[27] Whatever their poetic value, the various creation stories are almost always very poor astronomy and no better as geography, describing the Earth, or even a specific bit of real estate, as being at the very center of the cosmos.

There is still a round hollowed stone, the presumed Omphalos of the universe, at the Greek temple at Ephesus.

Astronomy did not begin with Copernicus, but it was the Copernican heliocentric universe which convincingly displaced Earth and its human master from geometric centrality in the solar system. The Roman Catholic Church was not amused and did not accept Copernicus' theory except as a "hypothesis" until around 1820.

Astronomers further removed Earth from the focus of creation by displacing the sun from its centrality among the stars. The geology of Charles Lyell and James Hutton in the early nineteenth century made the six millennia of our mythical history float out into the vast expanse of time that preceded it so that neither our place nor our time was unique.[28] None of this was pleasant for those trying to hold on to an anthropocentric world with an intelligent God that took a personal interest in day to day events.

It was, however, still possible to say that even if the sun was a star among stars, the Earth a planet among planets, and this year just the last in perhaps an infinity of years, nevertheless humans had a special place in creation. But when evolutionary theory inferred that humans were animals among animals, the last true uniqueness of our species was gone. In the Book of Job God permits the destruction of Job's wife, family, and wealth, but Job remains faithful. Satan then suggests that if his body were to be afflicted, this would shake him so badly that he would curse God. Of course Job avoided Satan's fundamentalist trap. For the Christian fundamentalists and their imitators, however, the Earth, time, the sun could all lose their uniqueness in creation, but howls of indignation arose when the self-image of mankind was touched.[29]

Did Evolution Really Happen?

But can the position of the fundamentalists really be refuted? It is not enough that evolution be supremely plausible. Did it happen that way, really? Aside from the mechanism's seeming plausible, what support is there?

Full documentation is beyond this book, but to be credible I must supply some examples of phenomena that make sense if evolution occurred and require more or less far-fetched hypotheses and ad hoc explanations if it did not. In short, simplicity is the criterion of the plausibility of evolutionary theory. In the absence of a plausible theory of evolution, we require that God have very curious properties, unacceptable to most theologians.

No sound empirical arguments have been advanced to refute evolution, and much evidence in favor of it is very difficult to explain away by any nonevolutionary mechanism. Unfortunately, the issue is confused by several dubious supporting arguments that are standard in elementary biology classes. These lead to confusion in the sense that they are compatible with either an evolutionary interpretation or an interpretation that supports deliberate design.

Here are three such arguments:

(1) The fact that organisms can be arranged in a treelike classification system is eminently reasonable if evolution has occurred but perhaps not awkward or meaningless if it has not. Imagine a mature tree. (A maple rather than a palm or a pine is better for this purpose, since the branches of a maple grow with almost complete independence of one another.) In the springtime it is possible to compare the length of green twigs at the tips of the branches and see that they have grown day by day. We do not see how the big branches covered with thick bark have originated but we know from looking at young trees that the thickest of branches began as a twig. We also see that there are many more twigs than there are thin branches and that each thick branch has grown many thinner ones. If you imagined it correctly, you can see that the tree grows by having many twigs, some of which develop twigs of their own and become stouter branches in the process. Most of the twigs die and drop to the ground along with the leaves in the fall. Occasionally larger branches also die.

In the same pattern, we visualize evolution as occurring by small differences among populations leading occasionally to new evolutionary twigs—separate and different populations and species. Most of these die out, but some give rise to other species or split into several species. The thickened branch may be called a genus, and if its daughter species have already produced new species it may be considered a higher category—a class or a phylum or even a kingdom, like that of plants or animals.

Just as the maple had one beginning and then branched out, so did organisms. Again, watch the metaphor carefully. The stout branches of a tree are actual living parts of the tree—they are different from the smaller branches, which are in turn different from the twigs. The stout branches that appear on diagrams of the tree of life are *categories* or groups of species and in that sense are intellectual constructs rather than living organisms or parts of organisms. That is, a genus does not exist independent of the species of which it is composed; when all the species within a genus become extinct, the genus itself is extinct. While this kind of argument is usually taken as supporting evolution, it is not really convincing. To an intelligent, educated fundamentalist it might seem equally plausible that the creator deliberately made the living world in such a way that it could be easily classified. If one were being asked to give up one's image of the created world and accept evolution in its stead, I can imagine that the treelike classification of organisms would not be convincing enough.

(2) Fossils support evolution. The age of rocks can be determined by dating on the basis of radioactive isotopes that have nothing to do with biology. The older the rocks, the more different are the fossils from living organisms. But even in the eighteenth century, when it became clear that fossils were leftover bits or impressions of plants and animals, some naturalists and geologists believed that a series of special creations had occurred, of which the one described in the Biblical Garden of Eden story was the successful culmination.

(3) The remarkable adaptations of particular species to particular circumstances are to be expected from evolution, but that is also what one would expect from a caring deity. Similarly, the evidence of comparative anatomy, showing that the same basic body plan is used with modifications within large groups of organisms, fits an evolutionary picture but equally well would be expected from a creator working out the variations on a theme. Consider that the works of any major artist involve repeated use and "evolution" of the same or similar themes.

In fact, all properties of living or fossil organisms are consistent with evolution's having occurred, but most properties are equally consistent with a fundamentalist position. I can think of only two exceptions. The first exception is the fact that evolution is still occurring. From what we know about organisms, constant miraculous intervention would be required to prevent it! This argument is not likely to convince the unconvinced, but I nevertheless find it attractive. Second, organisms are imperfect. This imperfection implies either that evolution occurred in essentially the way now accepted by most biologists or that the creator had a sense of humor or else was somewhat incompetent. For example, why are there different organisms filling very similar roles in different locations on Earth, and why do they have similar but not identical properties?[30] Why do some of these organisms seem poorly equipped for their role?

For example, I will choose just one family out of very many possible ones. Woodpeckers on the major continents are equipped with stout augerlike bills, with which they gouge holes in wood, and long sticky tongues, which can probe the holes and retrieve insects. On the Galapagos Islands there are no proper woodpeckers. There is, however, a finch—a member of a family of birds which in most places in the world eats primarily seeds and free-roaming insects—which labors as a jury-rigged woodpecker. It has a small but serviceable augerlike bill to make holes, but it lacks the long sticky tongue. However, it carries, grasped in its claw, a long cactus thorn with which it probes the holes, piercing and withdrawing insects, which it then eats off the thorn, much as a patron in the yakitori restaurant pulls chicken bits off bamboo skewers with his teeth.

Now, if an intelligent deity wanted to put woodpeckers onto the Galapagos Islands, why didn't he use a proper woodpecker? Conversely, if for some reason we can't fathom, this kind of deity wanted to have the finch do the job of a woodpecker, why wasn't the poor little bird given the right equipment? Since neither of these routes was followed, it seems far more plausible to believe that the rather hit-and-miss approximation is the result of evolution rather than that the deity was pathetically incompetent.

An even more remote possibility is that the Galapagos woodpecker finch was a kind of oversight in creation, but this scenario is rendered unlikely not by the rest of the world's being perfect (George Burns, portraying God in a film of that name, noted other imperfections, like the avocado seed's being too large) but by the fact that another weird, jury-rigged approximation to a woodpecker lived in New Zealand until it became extinct around a hundred years ago.[31] Finally, an alternative which is perfectly acceptable to me but seems utterly wrong to most fundamentalists is that the deity is not only intelligent in the human sense but has a sneaky sense of humor.

This curious geographic distribution of species of different apparent competences playing closely similar roles in different places may have forced both Darwin and Wallace to seriously consider a mechanism for evolution that did not require the ongoing attention of an intelligent creator. People who do research in evolution are scientists—they are neither philosophers nor theologians nor antitheologians nor revolutionaries, except perhaps in their extracurricular activities or hobby. Evolutionary biology is a proper and serious scientific discipline, with refereed journals, meetings, textbooks. Evolution draws on the findings from most of the rest of biology, including anatomy, physiology, ecology, and genetics, and these fields in turn gain meaning when viewed in the light of evolution. Except for its clash with fundamentalism, the theory of evolution has as little to say about religion as the theories of Newton did. But because the Hebrew Biblical text mentions the origins of organisms but does not develop an equally naive theory of physics, Darwin is seen as an enemy of orthodoxy, while Newton gets off scot free! This seems unfair. The only just alternative seems to be to abandon fundamentalism.

While I have no explicit documentation, I do have the impression that some leading evolutionary biologists passed through a childhood and early adolescence during which they were fundamentalists, and therefore their initiation into evolutionary theory became part of their maturation and liberation process. Also, a vocal subset of fundamentalists have been "born again," which in their case may mean that they lost faith in the secular intellectual processes. I suspect an element of zealotry against an old delusion on both sides in the trading of evolutionary and fundamentalist polemics. The conflict with fundamentalists is a curiously unintellectual battle, but it had to be discussed because of its currency and political importance.

Some More Sophisticated Problems with Evolution

In this chapter I have several times used the word "explain" but not "predict." This was deliberate. Evolutionary theory is explanatory, but it is not predictive in the way that astronomy is predictive. Astronomy predicts the date of eclipses and occultations. Evolution permits us to say "Of course!" after it has finished explaining some particular trait of an organism, but it does not generally predict what kind of organisms will appear tomorrow or in a hundred, thousand, or million years.

In one of its aspects, the theory of evolution does have the same kind of predictive power as the scientific parts of history or linguistics. All three can sometimes predict the properties that will someday be discovered about a past event. For example, the chromosome patterns of fruitflies permitted the prediction that there must have once existed, or that there still exist, flies with chromosome patterns that had not yet been found. These predictions were confirmed by new discoveries of living flies with just the predicted chromosome patterns.[32] Similarly, historians can predict that a particular document or person must have once existed and then later find direct documentary evidence for that existence, or the document itself.

Both history and evolution have a harder time dealing with the future. Both are concerned with complex systems, in which any development of one part can alter the future developments of other parts, in which small events, like the shooting of a not too bright Grand Duke, a cosmic ray hitting just the right gene in just the right way, or an unusually heavy frost at the moment when a rare plant is flowering, can set up important chains of responses that might not otherwise have occurred and certainly would not have occurred in just the same way. In fact "chance" events are important in evolution at several levels.[33]

All of the uncertainties of the environment over a full range of temporal and spatial scales impinge on any population. Rare largescale events like ice ages, volcanic eruptions, and enormous floods and fires obviously impinge on organisms. More locally, whether or not it will rain, whether or not some food plant will grow well in some particular time or place, whether or not some particular organism will meet a predator or break its leg tripping over a rock—

all of these may make an evolutionary difference. What if the organism which by chance broke its leg was the only carrier of what might have been a rare and valuable mutation? (Of course, there is a chance that the mutation itself might occur again.)

Chance also enters into the processes of sperm and egg formation and combination. While my daughter got half of her chromosomes from me and half from my wife, the egg that produces her child may have none of my chromosomes, all of the chromosomes that my daughter received from me, or some number in between. Obvious chance elements enter into my daughter's decision about who she will marry, and chance determines which sperm will fertilize her egg.

On the level of genes, generally genes do not change from generation to generation, but any gene is subject to the possibility of being struck by an environmental accident and mutating. Which gene will mutate, and exactly what its properties will be after it has mutated, are also chance events.

The ability to determine the rates at which chance events occur depends on having a large number of cases. No meaningful probabilities can be calculated for unique occurrences. Both historians and evolutionary theorists can tentatively state the probabilities, or more easily the improbabilities, of the future. Turtles will not grow wings. Flying fish will not develop jet engines. We will not be able to easily stop new venereal diseases like AIDS, but they will arise from time to time. It is impressive how few historians and experts on Eastern Europe managed to predict the deliquescing of communist control of Eastern Europe and the opening of the Berlin wall. The evolutionary theorists perhaps do better with the future than do historians.

We are accustomed to having science presented in mathematical form because of the history of physics. The Newtonian revolution in physics and astronomy was based on the discovery or, if you prefer, invention of novel procedures of mathematics. Newtonian and pre-Newtonian physics hinged on initially simplifying problems so that mathematics could be used. Because the importance of the theory of evolution is its power to explain organisms in all their complexity, a fully mathematical statement of all aspects of evolutionary theory is impossible.

The theory of evolution, as it was developed in the nineteenth century and as it continues until today, is mainly in verbal form. (The subarea of population genetics is a notable exception.) Scientific statements in verbal form tend to melt into metaphors. Since the bulk of evolution is verbal, and since evolution explains humanity as well as other organisms, there is an almost irresistible desire to somehow imagine that one's philosophical predilections arise from evolutionary causes and are not just wishful thinking. This desire results in perennial and unending discussion by biologists and philosophers, who should know better, of such matters as whether or not ethics can be derived from evolutionary laws, what the meaning of altruism is, whether or not natural selection "is creative," and so on and damn well on.

Todes has demonstrated that the phrase "struggle for existence" used by Darwin and labeled by him as a metaphor to express the consequences of the interactions of organisms with both their environment and with one another was repulsive to Russian biologists of the nineteenth century, while it was apparently neutral or helpful to Europeans and Americans. The metaphorical meanings assigned to aspects of evolutionary theory, even by scientists, vary with the social context, it seems—precisely what is not supposed to happen in science.[34]

Evolutionary theory also has an unfortunate tendency to develop polemical quasi-revolutions in very small subareas, which then are acclaimed as centrally important, perhaps only because they are controversial. In a style reminiscent of modern art criticism, and even more of the decretals and disputations that filled medieval libraries, many large volumes aggregate around even small controversies. I have omitted all of this week's controversies on the assumption that if they were included this book would rapidly become dated. I recommend Richard Dawkins' *Blind Watchmaker* for a spirited, but generally orthodox, discussion of these heretical schools.[35]

We must conclude that the theory of evolution is a scientific masterpiece, with powerful explanatory power. Because it deals with the complete assemblage of organisms, without simplifying the problem so as to make it tractable, evolutionists have succeeded in explaining great areas of nature in a rather general way, but have not developed the ability to predict, except for highly simplified subsystems.

In Chapter 3 I suggested that those areas of work in which simplification and complication are important are also characterized by a kind of playfulness. I think this has been made apparent in the discussion of religious doctrines, art, and most of science. In all of these cases, while there may have been the most earnest commit ments to the work, there were, nevertheless, rewards associated with maintaining the kind of objectivity and distance that charac- terize a player in a game. In the particular investigations of plants and animals that have entered into the theory of evolution as we now understand it, the same playfulness can be found. But when we ourselves were among the objects being studied, we seemed to have lost our distance from the subject matter. With detachment gone, intellectual work begins to lose some of its playfulness.

In the next chapter we will consider how the ideas of simplicity and complexity are used when people embrace the idea that they themselves must be made more simple or more complex.

9

Virtue and the Simple Life

The complexity of some problems may be so overwhelming that they must be deliberately simplified before work can proceed or even be thought about. This simplification replaces raw reality with a more or less elaborate intellectual game.

However, there are times, places, and attitudes that make simplicity and complexity much too serious and inclusive to fit within any game arena. Quite aside from the world of intellectual work, day-to-day living occurs in a world in which we are not free to decide the simplicity level at which we choose to play. We are pieces in the game rather than the controllers of play (Chapter 3).

Sometimes simplification of daily life is imposed by poverty or disaster. Sometimes simplicity is sought out as something virtuous or desirable. Simon Schama has written an elegant intellectual history of the Netherlands, from the beginnings of the independence war with Spain to the present. To a large degree this is an intellectual history of the desire for political and social simplicity. David Shi has focused on social and political simplification movements in America. These were invaluable background sources.[1]

Of course intellectual simplification is involved in production of any "self-image" (Chapter 1), but it does not necessarily involve verbal formulation. Verbalizations about who and what we are and how life should be lived in our everyday world are referred to as our "philosophy of life." There is a profound distinction between philosophical analysis and the development of a "philosophy of life." The first is an academic exercise for professionals. The second is generally done by amateurs. An analysis of simplicity and complexity by professional philosophers is in Chapter 10.

182

This chapter will look at the strengths, weaknesses, and problems of the "simple life" and its recurrent opposite—the life of deliberate ostentation and luxurious complexity. The simple life is generally seen as virtuous. More complex life, while not necessarily seen as vicious, requires more defense. Philosophies of life are often summarized in proverbs, songs, regulations, and mottoes. The richness and fuzziness of reality are very often conveniently encapsulated in assertions which become either hymns or platitudes, for example, "The simple things are best!" Enough truisms strung together either become a litany or a joke or some combination which may even be considered art. Mottoes may present some truths but also permit hiding of others.

There is a Shaker hymn beginning, "Tis a gift to be simple, 'tis a gift to be free." Its lovely tune is the theme of Copland's "Appalachian Spring." It also is an encapsulation of Shaker philosophy, but there is much more to the history of Shakerism and similar innovative ways of life than a charming motto.

While there are proverbs and mottoes in favor of ostentation, they usually do not make claims to divine inspiration or high moral value. "If you got it flaunt it!" "You're never too thin or too rich." Mottoes favoring ostentatious, or useless, consumption are a main product of the advertising industry. "Wouldn't you rather have a Buick?" "Just for the taste of it—Diet Coke!" (Check today's newspaper for a dozen more examples!)

In many parts of life we can discern a seeming reciprocity between simplifying one aspect of any problem and complicating another. Marriage simplifies some of the emotional problems that haunt single adults but in exchange sets different problems of mutual interaction, financial responsibility, relationships, and parenthood. Technology is both a simplifier and a complicator. A house full of electrical conveniences generates complexities of maintenance, repair, and expenditures unknown to the labor-intensive homes of fifty years ago.

One reasonable interpretation that an extraterrestrial observer might make of the United States—assuming that he is equipped with a powerful telescope and no further information—is that it is inhabited by noble automobiles served by human serfs. People could be seen washing, polishing, and feeding middle-aged cars, producing new young cars, burying the old ones, and using a major share of their income from other labor to support the cars that own them. The people keep pathways open for the cars, rearrange or even destroy their own burrows for the convenience of the cars, and sometimes lavish more care on the cars than on their own pups. Masterless humans (that is, people who do not serve a car) are old, crippled, poorly housed and fed, or immature. An alternative theory might be that television sets are really the main inhabitants of America. If this is funny, it is because it has some element of reality. Humor aside, a central object of personal convenience in America is the automobile. I did not learn to drive until I was thirty-five. I keenly remember the sense of avian freedom provided by my first automobile. Trips that once filled hours took minutes. I was free to decide when and where to go, which is impossible if we use trains or buses. Of course, almost all aspects of getting around are simplified for the individual by the presence of automobiles, but cars influence many aspects of life.

Cars generate problems. Aside from the expenses, licenses, and insurance forms, which are matters for the individual, the community must provide the car with smooth roads and must respond to the disadvantages and dangers associated with the use of automobiles as the primary transportation system. Some of these are obvious—air pollution, noise, accidents. Also, private automobiles destroy public transportation, with resultant difficulties in moving those that cannot drive their own cars, like school children, old people, the physically or neurologically impaired, and the deeply impoverished. The automobile obviously has ramifications for urban and regional planning, so that shopping malls that can only be reached by automobiles replace city streets as gathering places. The abandoned streets become frightening, or even dangerous, for the anachronistic pedestrians. There are also less immediately obvious consequences in areas as removed as wildlife conservation and global climate change.[2]

Despite their centrality and importance, automobiles are designed as commercial art objects. The greatest marketing coup of the twentieth century may have been to relate the automobile to sex.[3] Smiling, well-muscled young people in tight clothing, draping themselves over cars, are a feature of auto shows. Fenders, grilles, and upholstery are designed with deliberate sexual innuendo. Even the words "in the back seat of a car" carry overtones quite unlike descriptions of other locations—say "in the living room" or "on the beach" or even "in the bedroom"—though sexuality may be demographically more important in these immobile locations.

The role of the automobile demonstrates that at least some of the complexity of our society does not seem very sensible. But every society is complex, if not in technology then in its art, costume, religion, or class structure.[4] In fact, the main trend in social history seems to be toward complexity. This is a consequence of many things but most obviously of the acceleration of technological change that has been going on since the prehistoric origins of "complex societies" (that is, societies with definable classes and occupational specialization).

Increase in complexity may arise from obvious technological advances, necessitated by changing ecological, social, and economic problems. But we have also seen that much of the complexity seems independent of any really practical cause. I have followed Huizinga in assigning some of it to playfulness, which I see as an important part of the development of our human self-image, which in turn seems central to human evolution itself (see Chapter 1). Within the limits of solving real problems, a certain amount of playfulness is always permitted. Complexity can be added just for fun, at some cost in material and effort, if it does not seriously hinder the achievement of practical objectives. But sometimes ostentation becomes a goal in itself without serious regard to function.

Nonfunctional ostentation is a stimulus for works of social commentary. One of the more charming examples was written at the beginning of this century by Thorsten Veblen, who attempted to provide a general theory of ostentation.[5] Veblen believed in rapid evolutionary changes due to inheritance of acquired characteristics, and he also believed in racial memory and in the existence of a blond, long-skulled master race ancestral to the European and American "ruling class." It is therefore necessary to regard him with circumspection, but he is fun to read and very relevant to our theme.

His central point was that powerful men engaged in conspicuous consumption to demonstrate their power and superiority. He traced an imaginary prehistory in which the members of the ruling class, by force of arms, acquired property, largely in the form of slaves. Once the slaves were available, it was unseemly for the master to perform the kind of productive work that was done by the slaves. In fact, being as unproductive as possible, in the most expensive and conspicuous way possible, became a mark of mastery. Veblen concluded from this "mythic history" that essentially all of style, art, and civilization are motivated by a desire to demonstrate as blatantly as possible the power of the ruling class to consume in useless ways the productive work of the lower, nonruling classes. He then rapidly updated his history to a satirical description of manners and mores of the American middle class at the turn of the century, which leaves almost nothing unstoned.

He noted that a gentleman's clothing—tubular trousers, starched linen, tall silk hat, patent leather boots—and their fetish for cleanliness were advertisements of the wearer's avoidance of productive labor of any sort and of the fact that he could absorb the labors of others in keeping himself well dressed. The amusements of the well-to-do male—hunting, drinking, fighting, gambling—also were completely free of the taint of productivity. The wives of these gentlemen were dressed in voluminous skirts and giant hats with plumes and were generally constrained from any kind of free and easy movements. Changeable fashions not only altered the pattern of women's clothing every year but also required differences in body shape, which were enforced by stayed corsets and other complex undergarments.

He saw middle- and upper-class wives as slaves of their men, acting as surrogate conspicuous consumers to demonstrate the wealth of their masters. The wealth and nonproductiveness of a man could be measured by the elaboration and inconvenience of the costume of his wife, and by her nonproductive extravagance of time, effort, and money. The relegation of charitable and community work to the women was seen as demonstrating the basic unimportance of this kind of activity except as another demonstration of the male's power and wealth. Universities also appeared to be a matter of conspicuous waste of resources, at least insofar as they focused on what he termed "the higher learning," which seemed to be pretty much everything other than engineering.

Veblen generally shared the optimism of the beginning of the twentieth century, which could certainly have been justified by contrasting the material comforts of middle-class Americans in 1800 with those of 1900. He prophesied further improvements. He saw some hope in the rise of technical education and the gradual elimination of class distinctions. He correctly predicted that women would vote and would shed their voluminous skirts, but he also predicted that the practice of daily shaving by men would once and for all be abandoned. He saw shaving as a conspicuous waste of time and effort and a mechanism for making a man appear more womanish and thereby even less productive. But almost a century has passed and much of the optimism has been bombed and gassed out of us. Also, we cannot accept Veblen's flawed evolutionary arguments for the origin and meaning of ostentation.

Without accepting the curious history and biology of Veblen, there is still food for thought at least in his explanation of ostentation's having the social role of demonstrating wealth and power. Perhaps ostentation is in economics something like turbulence in hydrodynamics. Water can flow smoothly in simple straight lines if it moves slowly enough but breaks into "fractal" swirls if a critical velocity is exceeded. Complexity and ostentatious consumption are recurrent phenomena associated with a flow of resources beyond the capacity of simpler needs to absorb them.

A recent newspaper account described a new coffee shop being built in a department store in the Japanese former imperial capital of Nara. In this cradle of Zen Buddhism the shop offers "Evian water . . . heated in a gold kettle, poured through Jamaican Blue Mountain Coffee, dusted with real gold and served in gold-plated Royal Minton china. A genuine Renoir painting will hang on the coffee shop wall." The price will be $350 per cup. The proprietor describes the service he will perform: "We hope to make their lives more colorful."[6]

Tastes and moralities differ. While I find contemporary examples of extreme ostentation obscene, I can appreciate the charm of vanished historical examples. Huizinga describes a fifteenth-century festival at the court of Philip the Good of Burgundy. Part of the fun consisted of table ornaments, including "a rigged and ornamented sailboat, a meadow surrounded by trees with a fountain and a statue of Saint Andrew . . . and, lastly, a church with an organ and singers whose songs alternated with the music of the orchestra of twenty eight persons, which was placed in a pie."[7] (Donald Trump has never made a large enough pie.) Many more examples are available, catering to many kinds of desires.

While vulgar display requires wealth, poverty and virtue are not the only reasons for simplicity. Certain kinds of simplicity can be as ostentatious as any stylish game for the rich. Consider the playing at milkmaid by Marie Antoinette, which involved the toy village of Hameau, complete with "thatched cottages, a dairy barn with perfumed Swiss cows, and a picturesque water mill. When the lavish life at court proved tiresome, the Queen and her retainers could adjourn to Hameau, don peasant garb, and enjoy the thera-peutic effects of milking cows and churning butter."[8] The most complex ostentation involves calling on all of the arts to adorn one another in contexts in which the maximum number of possible appetites are appealed to simultaneously, even in the satisfaction of a simple need or desire. The playing of Parcheesi by the Mogul Emperor Ahkbar on the gigantic red stone playing field at Fatuhpar Sikree using slave girls as tokens, the reported feasts of peacocks' tongues in decadent Rome, the sex-laced gluttony of Diamond Jim Brady and Lillian Russell, are legendary examples of the deliberate elaboration of simple activities.

Riches, even very temporary riches, may beget complexity, or at least ostentation. In the Brazilian city of Manaus, 2,000 miles up the Amazon, when the demand for rubber tires for automobiles was just building and the only source was wild rubber trees, there was a period of enormous prosperity. It ended so abruptly in 1910, when the first Malaysian rubber plantations began to produce, that there are still half-finished buildings in the town with 1910 on their cornerstones. But at its crest the life of the town of Manaus must have been wonderful to see. Today the giant opera house still sits in the middle of the town, with its heroic half-nude plaster ladies around the walls, surrounded by its patterned black-and-white-tiled sidewalk that rivals that of Rio de Janeiro. In the lush days the streets around the opera house were paved with cork, lest the sound of wagon wheels and horses' hooves should disturb the keen esthetic senses of the planters and their ladies while they attended the opera. Similar stories can be told of the gold and silver mining towns of the western United States.

Attempts to Curb Ostentation

The ostentation of contemporary criminals, whose wealth may be very great but whose mechanisms for using it are culturally restricted, is obvious on downtown American streets. The use of gold jewelry by American high school students with after-school jobs not in drug stores but rather in drug trade, the sniffing of cocaine dust through rolled hundred dollar bills by America's glitz plutocracy, and the fancy dress of street pimps are current, familiar, and deeply disturbing examples of unnecessary and unpleasant complexities that may be attributed to excessive wealth suddenly flowing in unaccustomed places.

Notice the self-righteous tone of the last paragraph. I assure you it was not intentional when I wrote it. Evidently I have my own prejudices. It serves as a fine example of the attitude of simplicity as virtuous and ostentatious complexity as sinful which over many cultures and several thousands of years has led to sumptuary laws, blue laws, and simplifying reforms of all sorts from new agricultural colonies to Graham crackers. I will return to these after considering less uplifting causes of simplification.

Disasters like wars, the AIDS epidemic in sub-Saharan Africa, the famines of Ethiopia and Sudan horribly simplify societies. Some disasters are caused by nature, some by human perversity. There is no sign that they are becoming less likely to recur. At least as measured by material goods, some people in each society lead simpler lives than others—usually, but not always, because of poverty as individuals or as members of a repressed class. To share the simple life of the poor has very little romantic appeal in modern America. At present there is a strong correlation in the United States, and I suspect worldwide, between poverty and susceptibility to death and disease.

Without making a sociological survey, I would make an experienced guess that people who feel themselves to be living in poverty not only resent the absence of what are considered necessities but also resent the indignity of a continuing lack of more luxurious amenities. It seems understandable that impoverished people may hoard their resources for the sake of a display of apparent wealth on important occasions. Consider expensive weddings that throw families in debt, fancy funerals, hired limousines at high school proms, fountains of nonvintage wine at low-class bar mitzvahs, wedding anniversaries, and sweet-sixteen parties. These are sadly patterned after journalistic accounts of the tawdry celebrations of the very wealthy—the parties by the Great Gatsbys of the day.

As early as Solon's constitution for Athens, sumptuary laws have regulated private consumption, purchases, and expenditures.[9] The variety of sumptuary laws and their detailed intrusion into what we now in America consider rather private, and in fact unimportant, matters is fascinating but curiously monotonous. How many guests may attend a wedding feast was regulated—fifty guests and six musicians in Amsterdam in 1665, but only thirty-two, plus strangers from out of town, in Nuremberg in 1485. The Nuremberg wedding feast could only include fruit, cheese, bread, and cheap wine. The value of engagement rings, the trimmings on costumes, what the bride could take to her new home—not more than three robes in Solon's Athens in the sixth century B.C.—were also regulated. No multicolored dresses in third-century A.D. Rome. Engagement rings were limited to those that cost less than ten gulden in fifteenth-century Germany. In fact, all the occasions which tempt one to show off and splurge were at one time or another, in some place or other, subject to sumptuary regulations.

The motivations differed with time and place. Often ostentation was equated with the kind of dangerous sin that can bring divine retribution. Sumptuary laws were often instituted or reinforced after floods or earthquakes or epidemics, and dire references to the fall of Rome, or other large objects, are often heard when discussing modern ostentation. While sumptuary laws were promulgated by the pre-Reformation clergy of Europe, the reformation zeal in enforced simplicity has permanently linked Calvinism with sumptuary laws and blue laws. Simon Schama repeatedly refers to the ongoing admonition of the Calvinist clergy against the ornateness of dress and furnishings of the wealthy.[10]

Most sumptuary laws provide a temptation to some kind of circumvention. A sinless way to enjoy wealth used by the Dutch upper class, at least during the seventeenth century, was to reduce the colors of their dress to essentially black and white—but the white was of lace or of the finest and most meticulously bleached and laundered linen, and the black was velvet, silk, or other extremely elegant fabric. Simplicity and cleanliness were raised to the level of ostentation.

Until the development of organic chemistry in the nineteenth century, brightly colored cloth was a great rarity. The kinds of brilliantly colored pigments that were used in painting, since time immemorial, were not the kind that made good dyes. For dying cloth, yellows, greens, browns (all reasonably subdued) were available from plant materials. Cochineal bugs were slaughtered in their millions to give a red. Blue came from indigo but was not colorfast—so that the indigo-cloth-wrapped Tauregs of North Africa had their faces stained as if by carbon paper and were known as the "blue men."

There was, however, one brilliant dye which, depending on detailed conditions, gave a color ranging from blue through various shades of purple and scarlet and was reasonably free from fading. This was the famous Tyrian purple derived from one of several species of snail. These snails are found in various places but were most heavily exploited on the Eastern Mediterranean shores of present-day Lebanon and Israel. The color was derived from the snails' digestive glands. The process of dying the wool involved oxidizing the digestive glands, which are initially yellow-green. Depending on the more or less uncontrollable effects of the oxidation, the actual color would vary from very dark purple through to violet and crimson and blue.

Purple cloth has been a particular object of concern in sumptuary legislation. Permission to use the color purple was an important privilege of rank in Rome and Byzantium and before. Emperors were born in a special palace of purple stones—so that it might be said of them that they were "porphyrogenitus"—born to the purple. There were severe penalties for use of purple cloth or even threads by unauthorized persons.[11]

Why there was such a remarkable focus of sumptuary legislation on the color purple is not completely obvious. Perhaps it was because the utility of cloth was in no way enhanced by this expensive coloring process, perhaps it was the unique brilliance of the color. Quite likely it was simply that purple cloth was expensive. Even today purple carries an overtone of sinful ostentation—even a red Cadillac seems more suitable than a purple one.

But this raises the question of the intentions of the sumptuary laws. Sumptuary laws were often designed to merely save money. By curbing ostentatious display, and particularly by limiting the amount of money that could be spent on a public occasion, the pressure on middle-income families trying to keep up with their economic betters would be relieved. As economic inequities decreased in Nuremberg, the pattern of the sumptuary laws shifted to maintenance of class distinction, rather than merely saving money. While it was, and is still, generally believed that simple esthetic taste is an indication of morality, it is a type of morality most becoming to the poor and to the lower classes.

Low-class status and poverty are strongly associated in American thought, but this is not universal. The sumptuary laws of Tokugawa Japan regulated the behavior of the merchant class, which was understood to be socially inferior to both the samurai and the peasants but as a matter of fact owned most of Japan's wealth. Despite their wealth they were, at least for a while, forbidden to wear ostentatious kimono. This restriction accounts for the elegant, eighteenth-century black kimono, with its brilliantly colored embroidery in the lining.

Generally self-appointed powerful aristocracies entitle themselves to greater flamboyance. The 1696 Nuremberg legislation decreed, among other things, that women of the "old nobility" ("Alten Adelichen Geschlechts") could wear caps of velvet, with sable or marten fur trimming, which at festivals could be ornamented with gold and a few, but not too many, pearls. Ladies of the respectable merchants', or second, class could not use gold or gold lace on their caps but could use gold mixed with silver. Ladies of the third class could use velvet caps with silver ornaments. Fourth-rank ladies could use "Tripp-velvet" caps but could not use marten fur trim and no metal ornaments. There were other regulations on other articles of dress, conveyances, style of weddings and funerals, and so on. Just because low-class people had money, they had no right to look as good or live as well as their betters. Shades of the bad days in the American South. When I lived in a small southern town in the 1950s I was told by blacks and whites that a freshly painted house owned by a black could be a dangerous provocation to racism.

Sumptuary laws have never died out. Iranian law under the

Ayatollahs regulates women's clothing for the sake of general morality and well-being. In America, given current interpretations of
the legal separation of church and state and the wording of our
constitutional amendments, laws are assumed to regulate interpersonal interactions rather than the morality of individual behavior.
But this is very recent indeed. Laws about gambling are sometimes
defended in much the same language as the defenses by the city
council of Nuremberg of their sumptuary laws, namely as forestall-
ing sinful expenditures. We usually justify drug and prostitution
laws on the basis of health rather than private morality. Most of us
no longer consider decolletage an invitation to divine retribution,
but many argue that private lotteries are dangerously sinful.

Tobacco is as useless as the color purple, but more addictive. To
smoke is a private activity, unless it occurs in a public place, when
it can be as offensive to others as consumption of raw garlic, with
much more dangerous consequences. On the other hand, since major
public costs are incurred for medical aid and since tobacco- and
alcohol-related illnesses and accidents fill almost half of our hospital
beds, neither tobacco nor alcohol are truly private problems.[12] In
short, while we may have outgrown pure sumptuary laws, they are
more than quaint anachronisms. They are to some degree models of
how public law meets private morality and "philosophies of life."

The Philosophy of Virtuous Simplicity

Those individuals and social groups who have preached, and even
adopted, simplicity as a virtue, without being forced to do so by
social or economic restrictions, are of particular interest. All through
recorded history some individuals have embarked on simple living
as part of their "philosophy of life," without waiting for the range
of intellectual, esthetic, or metaphysical problems associated with
simplicity to be solved.

Some have embraced simplicity quietly and privately. Others
imposed simplicity on their immediate families, perhaps with curious
results for the lives of their children, although I know of no system-
atic study. The motivations may involve supposedly factual considerations of health, morality, politics, or religion. By and large we
don't know much about these private simplifiers. They are often

dismissed as cranks. Some adherents of simplicity do so in a more social way and "witness" their simplicity (to use the Quaker term).

The young lawyer returning to India from Durban, South Africa, giving up his high collars and neckties to wear a dohti and sit on the ground with a little wheel spinning cotton threads as a Mahatma; Diogenes the Cynic sitting in his tub and trading wisecracks with Alexander the Great; Thoreau camping out in the backyard of Concord, Massachusetts; John the Baptist moving into the desert to eat grasshoppers; Albert Schweitzer playing Bach in a bush hospital, with a gazelle chained by the neck outside his tent; the Egyptian Desert Fathers, living as hermits, without even their books but within easy walking distance of the market, were all "witnessing." They were using the simple life as a comment on the complexity, and hence sinfulness, of merely accepting life as it comes. In fact, a minimalist philosophy of life can illuminate the usual complexity of living the way minimalist art illuminates the rest of art.

Ostentatious simplicity can be an important part of a political program. Thoreau's going to jail was a very general kind of statement. Mahatma Gandhi had a much more complete program. The proponents of sumptuary laws usually are attempting to either return to past virtues or prevent the decay of present virtue. But they do so from the position of identification with the current authority of the society.

Those who bear witness to simplicity with their lives may articulate similar desires to return to an imagined past but do so in a mode that leads to social revolution or at least to the construction of islands of rejection of current mores and of current power structure, standing as a rebuke to the complexity of the world around them. They may create sects, minor religious revolutions.

For better or for worse, certain aspects of European religious certainty burned to ashes in the Lutheran Reformation. In its wake, ways had to be found for finding answers to contentious questions without God-ordained priestly authorities or a clear rabbinate with the authority of tradition. No one was authoritative in the eyes of all the participants in disputes. The Reformation required that the assumption that the world and mankind were "in God" be shifted to the assumption that God, for practical purposes, was "in the world," or even coextensive with the world, as in the pantheism of Spinoza and of the Romantics, or that God was in each individual human, as in the doctrines of many Protestant sects and early scientists. This made it more difficult for the authority figures to provide clear answers, since they became subject to obvious questions or accusations: "By whose authority is that said?" "That is not the voice of God, but is either your misunderstanding, or perhaps the voice of the Devil."[13] For individuals to prepare justice, ethics, and theology from scratch was a new problem, perhaps never having arisen before in all of European and American history.

The civilization that grew out of the Reformation developed a new theology in which God within each person provides direct inspiration for words and action. This is seen in a very pure form at Quaker meetings in which it is assumed in principle that God directly inspires speakers. Several American denominations in the early nineteenth and late eighteenth century forbade the writing of sermons on the theory that the pastor delivering an unprepared sermon was "guided by the Spirit's energy" to speak "the truth in the demonstration of its might."[14] The present-day Church of God, the Mormon Church, some branches of the Baptist Church, and most of the native fundamentalist Protestant sects began with the assumption that the voice of God can be found by seeking within one's heart.

In another branch of the river of the Reformation, concern for revealed religion was replaced by a concern for natural law, by Adam Smith's concern for "sympathetic sentiment," which implied that the truth is found within the human heart, soul, or mind and that this is somehow a property of being human.[15]

Sometimes, and in some places, the founding of new sects becomes almost a style, as on the American frontier 200 years ago. Among the most curious of these sects was Shakerism, which is of particular interest as a relatively recent religious community focusing on simplicity at many levels. They created "simple" masterpieces of furniture and implement design, along with nightmares of sexual frustration and hysteria.[16] Shakerism appeared only in the late eighteenth century and voluntarily became extinct in the mid twentieth century, a period spanning the establishment of the United States, the Industrial Revolution, and the age of the automobile and radio. Its surviving artifacts and buildings are of primary importance in the context of museums, not of current religion.

Amish and Hutterites, some Catholic monastic orders, some of the hasidim, Amerindians, bushmen, fakirs, kibbutzniks, and other contemporary communities in different ways and to different degrees have decided on a simple life, by which most of them mean rejection of the current fads, frontiers of fashion, clever innovations, and wealth and status symbols. In the past there were, in addition to the Shakers, Waldensians, Albigensians, Essenes, Prophets—to mention just a few—all of whom in one way or another sought out simplicity and then vanished.

For these groups the issue was, or is, a moral and political one at least as much as an intellectual one. Simple is good, complicated is bad, although it is often the case that necessity and choice inter-twine for those taking pride in their simplicity and poverty and sometimes even pride in their ignorance. Advocates of the simple life may come from all social and intellectual classes, but there is always the danger that the simplifiers may discard things of great value as vanities and useless complications. "Let us make a great simplification and then the world shall begin again" is a terribly dangerous assertion.[17]

Self-Chosen People

There are two basic styles of political programs for simplicity. First and least dramatic are the reformers and would-be reformers, who may even take charge of a government by peaceful or revolutionary means. In retrospect they are simply another phase in a country's history. Orwell's noting the resemblance between the pigs and the farmers in *Animal Farm* is an important generalization. The dis-tinction between considering Cromwell as the religiously inspired protector and as a rather dour dictator between kings is much less evident than it once was.

More dramatic are those who establish new nations, either by splitting territories or by emigration. Schama focused on the Dutch giving birth to their own nationality during their rebellion against the Spanish monarchy, in much the same way as the United States defined a new nationality in fighting Britain. The many small nations that burst from the belly of the Austro-Hungarian Empire early in this century carefully invented national costumes, anthems, and mythic histories and emphasized their own linguistic distinctions from their neighbors. Perhaps most of what we now call nations were born by proclaiming their identity and then creating a self-image. Is it possible that the distinction between a revolution and a civil war is that the first is redefining identity as well as disputing other political and social issues, while the second is attempting reformation of an existing identity? I expect that the decay of Russian hegemony will demonstrate aspects of both.

Puritans of Massachusetts, Quakers of Pennsylvania, Voertrekkers of South Africa, and the Zionist Halutzim of Israel combined construction of new identities with moving into "new" lands. In each case, the image of an empty land in which to build was simplistic and dangerous. Also in every case, the idealistic goals of the pioneers were corrupted in one way or another. William Penn's son became an Episcopalian, and children of socialist kibbutzim leave to study finance (or even enter the rabbinate). Often, the later immigrants to a new colony come for reasons unrelated to the founders' goals. Also, to the degree that the new society is materially successful, the simplicity and austerity of the founders seem less relevant.

The Puritans and Quakers left England to build godly simple republics. Hard work and simple wants were seen as adjuncts to high moral standards, freedom from want, and access to heaven. There is, however, an instability in the temporal continuity of the institutions established by industrious searchers for the simple life. While they assumed an association between riches, complexity, and sin on the one hand and poverty, simplicity, and virtue on the other, they made no provision for the fact that simplicity and hard work may generate wealth, which threatens simplicity. To want very little when wants are unrestricted by poverty seems difficult. Wealth may even make heaven seem less important.

Unfinished Real Questions

I do not feel that much interest or affection for those that look for the complicated life. As Shi points out, there has been an association in America for the past several centuries among some kind of pursuit of the simple life, liberal politics, and academic status. It is not accidental that my daughter is a handweaver, my wife a hand-spinner, one son a violin maker, and I an academic ecologist and third-generation Democrat. We raise a good share of our fruits and vegetables, and I sail a boat and use the outboard as seldom as possible. Perhaps I would have been unhappy if my children had worked for their MBA, but I'm not sure.

Despite my own family's proclivities, I cannot blindly accept the idea that the simple life is obviously virtuous and a complex life is not. Such an attitude would smell of fundamentalism, to which I have an allergy. I must therefore consider what determines the limits to possible simplicity and complexity in living. Obviously one limit is material. You can't waste it if you don't have it, and there is no virtue in lack of possessions or abundance of possessions in them-selves. At the limit, "excessively complex" means that it becomes difficult for people in the society to construct a reasonably consistent self-image. Drastic reorganization must then occur. Individuals and small groups that cannot form a satisfactory self-image are ripe for madness. Societies in which formation of a satisfactory self-image is actually impossible for a large fraction of the population are ripe for some kind of revolution, invasion, or both. This is not mere rhetoric.

Between individual madness and full-scale revolution lies the possibility for single individuals, families, or larger social groups to reconstruct their own social systems in a search for a "better" world. This almost always involves a shift along the simplicity–complexity axis. As we have seen repeatedly, which way one is moving on this axis is subject to differences of definition, but an internally consis-tent self-image must be maintained or created. For example, the "Hagarism Hypothesis" of the development of Islam (see Chapter 2) explains the differences between the story of Islam in Syria and Iran in terms of the multiple internal inconsistencies of the possible self-image available in Byzantine Syria as contrasted with the polit-ically weak but still internally consistent pattern of Zoroastrian Iran. In the first case, the conversion to Islam involved a revisionist reconstruction of Syrian history, so that Syrians of many kinds claimed Arabic identity. In the second case the conversion was to the religion of Islam, producing Iranian rather than Arab Muslims.

The curiously passive but very rapid deliquescing of the Communist system of Europe, which is happening while I write, will, I expect, eventually be seen as another example of decay from internal con- fusion in self-image.

There are other unresolved questions about the "simple life." If I can maintain intellectual and emotional simplicity by using won- derfully complicated devices such as the word-processor with which I wrote that sentence, am I living the "simple life," or does living the simple life demand that I learn complex skills in the use of hand tools? But also, what is virtuous about the simple life? Is elaboration of daily life a sin, a privilege, or a perversion? Is simplification at war with complication, and is this an aspect of the war between the saved and the damned? Is the "simple life" virtuous, silly, or simply a result of romanticizing about poverty? It has been repeat- edly implied or stated outright that the simple life is an imitation of Adam and Eve before the expulsion from the garden. Is that where the imagery of the conjunction of simplicity and virtue came from?

All of these questions have at various times and places been a matter of life and death decisions. The fact that people who make these decisions have never analyzed simplicity or complexity very deeply does not negate the seriousness of the decisions themselves. Clearly we are no longer discussing the playing of games. We will in the next chapter see what the philosophers, most of whom are very serious persons, can tell us about real-life simplicity.

10

Masters of Reality

By focusing on examples rather than definitions I have permitted words to float like plankton in a vernacular sea. This had the advantage of allowing us to see "simplicity" and "complexity" in their natural habitat, unrestricted in their behavior and free to attach themselves in any way that they would. We have watched them change meaning and role. Now we must formalize this natural history. I must consider what simplicity really means. Is simplicity completely context-dependent, or is there really a common set of phenomena underlying all that we have seen, or have we been looking at an arbitrary collection of ideas hiding under a few common words? Are the problems of simplicity and complexity really only related to how we represent the world and what we want to accomplish in it, or can they be productively discussed in a more general sense? Are some things really more complex than others?

I hope you worry when I use the word "really." Nevertheless, we must face the fact that most people feel nervous as players of intellectual games. They want to be *really* serious. Meeting this desire is neither simple, easy, nor even completely possible. We must also understand that simplicity and complexity are sometimes more substantive than merely matters of verbiage.

Definitions

Dictionary definitions help in understanding how words relate to reality, but not very much. They are also simplifications, expressing the interests, and often the prejudices, or humor, of their author. Consider Samuel Johnson's famous definition of "oats" as "A grain, which in England is generally given to horses, but in Scotland supports the people." The 1947 *Webster's Dictionary* contains: "Har-vard crimson . . . A color, red in hue, of high saturation and low brilliance." The same dictionary defines "Yale Blue . . . A color, reddish-blue in hue, of high saturation and low brilliance."[1] Names of nations, races, political parties often have prejudicial definitions which are not at all funny.

"Simple" does have a straightforward, if archaic, meaning as a noun: "Any medical plant or the medicine extracted from it: from the former supposition that each single herb was or provided a specific for some disease. 'A great knowledge of simples for the cure of disease is popularly attributed to the Indian'. F.Parkin, 1867."[2] The verb "to simple," meaning to gather simples, is defined by Johnson. It is however more difficult to define "simple" in its deeper meaning than it is to define "oats."

Take one attempt: "Everything confronting man in experience admits of composition of some kind; thus one's procedure in arriving at the notion of simplicity is necessarily nugatory. The concept itself signifies a negation of composition: a simple thing is something that lacks parts or really distinct elements. Simplicity likewise implies indivisibility, since only composites admit of dividing." So far so good. "Simple" is here taken as an adjective like "green" or "pretty." "Simplicity" is the state of being simple, like objects can be pretty or green. There is room for discussion about whether or not absence of parts is the only useful hallmark of simplicity, but we will come back to that.

The definition then continues to clarify the idea and to define a special usage. "In the created order the complex is more perfect than the simple, as man is more perfect than a stone. This, however, is simplicity of imperfection or lack of being. The Simplicity of God consists of being all perfection."[3] This reiterates an aspect of the idea of the previous chapter that the simple and the virtuous are associated. But this sets up an association between two abstract nouns. The associated nouns have no more practical meaning than either one singly.

Many abstract nouns, like "the simple," "the complex," and "the minimal," tend to resist doing any hard work in practical discussions. However valuable they may be in metaphysical contexts and as poetic images (see below), nouns representing essences, qualities, and absolutes, like choruses in plays, may inform the audience but the plot does not proceed at its usual rate while they speak. It either stands still or jumps. The idea of absolute "communism" had somewhat the same role in Lenin's Soviet Union as did "simplicity" in a Shaker family.

Adjectives do more work. But even as an adjective, "simpler" is usually asked to carry moral, intellectual, and metaphysical burdens that we would not require of most other adjectives, which is the rationale for this volume. There would have been very little point in attempting to write a book about "green," for example, except perhaps within the context of paint chemistry or physiological botany.

For most of this chapter, simplicity will be considered on a comparative, operational basis—something is simpler than something else, or something has a particular relative simplicity value as measured by some specified procedure.

Determination of relative simplicity requires ranking, which, in turn, requires some classification system so that we have a series of objects which are in some sense comparable. To rank or classify for the sake of ranking and classification seems empty and in the tradition of Teutonic academics. Usually the act of ranking is undertaken in the context of proposed further decisions. Ranking of students facilitates decisions about scholarships. Soldiers may be ranked hierarchically for purposes of transmission of orders. Objects for sale may be ranked to aid pricing, and so on.

Useful ranking requires classification. We can think of better or worse apples or oranges, but the proverbial prohibition against comparing apples and oranges is misleading. We may indeed want to compare apples and oranges for some purpose (say, their relative value as fruit salad ingredients or their vitamin C content per dollar expended). In order to do this we must have defined some broader category to contain them (fruit or food). To attempt to rank students and oranges in one sequence is not of obvious use but is by no means out of the question.

To determine the relative simplicity of something is to compare an object, either ponderable (say a machine) or theoretical (say an equation), with another real or hypothetical object which would be ranked before or after the object in complexity according to some agreed upon procedure. The objects ranked would all have to be members of some common group. What aspects of objects ought to be considered in classification or ranking will depend on our goals.

Sometimes we mean by simplicity something like "operational effectiveness" or "ease of manipulation." Some objects are easier to make, describe, repair, grow, or replace than others. I can imagine different criteria of simplicity defined for the same set of objects which would produce different rank orders. For example, I wear an old automatic Omega wristwatch (a gift from my father-in-law thirty years ago). It began losing time this year, and I found that it was expensive to repair and that watches of much greater structural complexity could be purchased at a relatively small price. These cheap complex watches cannot be repaired at all. They are as disposable as paper diapers.

I can defend many separate ways of ranking the simplicity of my old, and old-fashioned, watch against that of a new one made in a more advanced fashion. Among the possible ranking criteria is the ease of performing a repair, which may be measured in terms of the prevalence of persons competent to do the job, or of years of training required to gain such competence, or of cost of performing the job once it is decided that it should be done. Since the new watches cannot be repaired for any price, my old one is simpler by default. A second ordering system might involve counting moving parts, after deciding how big a part must be before it is counted (to avoid problems with, for example, watches that run on radioactive atomic decay). A third involves ranking the complexity of scientific theories underlying objects. In this case, Newtonian theory is suitable for my own watch, but complex circuitry, at the very least, is involved in making the new watches. A fourth scale is a personal one relating to some kind of internal narrative of my life.

Therefore, my old watch is simpler than the new ones from the standpoint of repair, and operates on simpler scientific principles. It is more complex from the standpoint of number of moving parts, and also from my personal standpoint—my relation to anonymous objects or persons being much simpler than my relation to things that are intertwined in my personal history. Cost has also been used as a measure, if not of complexity, of lack of simplicity (see Chapter 9). By that standard my old watch is less simple than a cheap new one, since the repair would cost more than a new watch.

There is obviously yet another standard related to the use of the watch. My old watch tells the approximate time and the exact date (assuming I reset it for February, September, April, June, and November, which information resides in a doggerel rhyme in my head). Newer watches will not only keep the day of the month straight but also remind me of birthdays and appointments and give precise time to fractions of a second; furthermore, I suspect from simplistic historical extrapolation that there will be models for sale by the time this book is printed that will also provide telephone area codes for all the world, birthstones for all the months, and a complete list of state flowers and state birds. The purpose of this complex of information would be to simplify my life.

The awkward richness of possibilities seems to shatter any possible coherent theory of simplicity, which doesn't seem reasonable. Also notice that while some of the criteria involve apparently objective procedures, they each ultimately rest on some intuitive feeling on my part for how to devise the scale. Have I woven such a densely complex web of examples that I can't find my way out of it? Are some of the ways of measuring relative simplicity of watches illegitimate? Should I perhaps now rank the ranking methods themselves in order of simplicity, and if I did so would I find some virtue in choosing the simplest? How do I decide on a proper method? Lacking personal expertise to solve this problem, where can I turn? Two professions claim to deal with this kind of problem: mathematics and philosophy.

Sometimes it becomes necessary to call on mathematicians or philosophers in almost the same spirit as calling on a plumber, dentist, or tax lawyer. At least one difference comes to mind, however. While it may not always be obvious what tax lawyers, dentists, or plumbers are actually doing, we usually can easily understand what they are doing it about.[4] Also, the problems dealt with by dentists, plumbers, and tax lawyers have an immediacy which lends an excitement to their work that is less obvious in the work of mathematicians and philosophers. Nevertheless, when intellectual problems are so difficult that commonsense thinking about them becomes impossible, mathematicians and philosophers are necessary. The practical need is to have the intellectual difficulty, or the problem, or both, either go away or become tractable.

Mathematicians carefully circumscribe interesting domains by choosing a set of axioms which define the world that they are going

to be concerned with for the moment. Not all problems lend themselves to this kind of neatness. A philosopher cannot reject a problem
because it is not philosophy in the same way that a mathematician
can reject a problem for not being mathematics.[5] On the other hand,
philosophers, aside from pure logicians, must deal at most with this
world. Some philosophers take as their mission facing any and all
intellectual problems and beating them into tractability, or at least
attempting to.

Since many philosophers concerned with simplicity and complexity draw heavily on the work of mathematicians, but the converse
is not as conspicuous, I will very briefly consider a few aspects of
what mathematicians worry about, as a preface to considering how
some philosophers have dealt with simplicity.

The work of mathematicians is to imagine new worlds and what
their regularities might be. In their choice of difficult intellectual
problems they can exclude questions that require physical apparatus
for collecting information or knowledge of political or social history
in their answers. In this sense they are free of the limitations placed
on empirical scientists, who are supposed to examine this world.[6]
(Computers manipulate but generally do not gather new information. Also, most pure mathematics doesn't require computers.) The
purpose of their activity is clarity and understanding only. While
they are free to make new worlds, they carefully avoid dealing with
the inconvenient parts of this one.[7]

Pure mathematics[8] is not undertaken to construct anything tangible, nor is it to provide purely esthetic pleasure, although it is not
free of esthetics and occasionally is helpful in solving engineering
or other operational problems. The definition of mathematical quality is in part esthetic, but conceptual difficulty is also part of the
standard. Mathematicians, like other researchers, prize novelty, and
easy-to-imagine kinds of mathematics have almost certainly been
explored already and are therefore uninteresting. Difficulty is not
the same as complexity. Simple solutions to difficult problems are
highly prized; complex solutions are less valued, but simplicity and
complexity have technical definitions we will return to. Difficult
mathematics does not necessarily involve difficult computations, and
in fact may not involve computations at all. Difficulty here involves
conceptual difficulty.

A slight venture into two examples may help clarify this and subsequent discussion. The first involves the idea of "dimensions" and the second that of "number." (I find that I develop a slight headache in rereading the next few pages. Nevertheless, I think it is important to include them.)[9]

What regularities might be expected in a world of a larger or smaller number of dimensions than the three dimensions of the builder or carpenter? Why should we care about such a curious space? Partially because the game we can play with it is intrinsically fascinating, partially because if we are willing to simplify our sense of space in this way then new empirical problems can be formulated and sometimes even solved. By accepting this we are intellectually free to think beyond our own pedestrian set of obvious dimensions.

This is a very important and practical idea. On one hand, it immediately recalls Plato's myth of the cave (see Chapter 1). What was wrong with the perceptual world of the poor prisoners in the cave was that they saw only the two-dimensional shadows of three-dimensional objects. In principle, by noting the changes in shapes of these shadows they might eventually have developed a three-dimensional theory of objects: but in the absence of a general notion of dimensions, that would have required great intellectual force indeed. We now have general techniques for considering how many dimensions are relevant to analysis of some sets of data. In a general sense, more complex things require more terms in their description than more simple things. The number of dimensions required to specify a situation depends on how much complexity we are willing to consider, and this may depend on what we want to do or say.

For example, in my laboratory some years ago we simultaneously measured the body size, reproductive rate, and food consumption of hydra of several species. To present the data as a table did not aid visualization; to present them as a straight line in two dimensions (say, food supply plotted against size) resulted in a fuzzy cloud of points; but if we considered a two-dimensional surface in a three-dimensional space (with the dimensions being size, budding rate, and food), then the data points fitted reasonably neatly, although not perfectly. The imperfections were the subject of further investigation, which made it clear that in addition to characterizing each animal by the three measurements of size, budding rate, and food quantity, we had to include food particle size in describing the state of the animal. This is obviously a fourth dimension, and with these four dimensions we could make evolutionary and ecological predictions that we could not make with only three.[10]

We had theoretical reasons to assign empirical meaning to these dimensions, but even if such an assignment cannot be made, it is still often useful to at least consider how many dimensions might be relevant and how they should be arranged in "space" in order to suggest ways of thinking about data. There is no necessary numerical limit to the number of dimensions involved. The more kinds of things we know about anything, the more likely we are to be able to deal with it in a realistic way. We need not be prisoners looking at shadows, but at the same time we must be careful how we make inferences about higher dimensions. In two or three dimensions, geometry works the way high school textbooks say it should, but it is different with more dimensions. How different, and what difference it might make, constitute possible questions for mathematicians.

So now we have identified at least two kinds of problems: "How can we say that something is simpler or more complex than something else?" and "What is the relation between dimensionality and complexity?" I will touch lightly on both of these.

Before objects can be ranked in order of simplicity we must somehow decide which sets of objects are sufficiently similar so that a common ranking has any meaning at all. To say that a set of objects is of the same sort, so that ranking, among other operations, becomes meaningful, is to say that in at least some of the dimensions of their multidimensional world they are closer together than the objects in some other possible sets.

For example, to be "near" has the meaning of physical proximity if we consider three ordinary spatial dimensions. It has the meaning of temporal proximity if we consider only the dimension of time. It has the meaning of historical association if we consider space and time—if someone or something was in existence at the same time as George Washington and was sufficiently close in space to Washington during that time, then they may have interacted. While these criteria eliminate almost all the objects that have ever existed, we would need more descriptive dimensions to distinguish between Washington's wife, dinner, horse, chair, and mother. Nevertheless, proximity in four-dimensional space can define the class of objects that George Washington could have touched. The same principle, involving different dimensions and larger numbers of dimensions, can be used for classifying other kinds of objects into groups.

The idea of dimension very quickly loses geometric identity, so that any world of objects, each of which requires some particular number of measurements, say k, to assign its properties, can be thought of as being in a k-dimensional "space," which should not be thought of as our geometric space, or any contortion of it. An intellectual simplification is being called for. Roughly, the dimensions of a space are measured by the number of numbers required to specify a location in it. In a space we have some rule for telling when points are more or less close together.

One important use of this sort of space is in the classification of organisms, which we required in the discussion of evolution (see Chapter 8). It is common to speak of some set of similar organisms resembling some other more or less closely. But organisms cannot be very satisfactorily identified by only three numbers. Many separate measurements can be made of the shapes of organisms. If similar measurements have been made on a set of organisms, we can think of each of the organisms as points in a "shape" space or "morphological" space. It may be useful to consider how far objects of this sort are from each other in that space.

What do "distance" and "useful" mean in this context? There are various possible answers to these questions. Distance in a one-, two-, or three-dimensional space—that is, in the spaces we can easily visualize[11]—has the property of being additive, so that to go from point A to point B and then to point C is a longer trip than going just from A to B. Also, shapes and locations can be defined in terms of distances, so that A, B, and C may actually be arranged so that C is very close to A and far from B, which would leave the addition of distances intact but raise the question of how wise it was to go from A to C by way of B. Any functions of the location of objects in higher dimensional space which preserve what seem to be reasonable analogs of what we mean by distance in three-dimensional space are called "distances," usually with some explanatory adjective attached.

Generally, "good" (that is useful) classification systems incorporate information in designating the categories, thereby reducing the additional information required for operations within a category. This constitutes an apparently "real" simplification, since all the information that holds for an entire category or nested set of cate- gories need not be repeated for each individual.

For example, a large group of animals share the property of having a backbone; they are called "vertebrates." A large subgroup of these, the "mammals," have hair and also can give milk. One group of mammals have properties that permit them to eat meat and are called "carnivores." Of these, one group walks up on their toes and has peculiar scissors-like dentition. These, the "felines," include several subgroups differing in such things as body size and general attitude. One of these is called *Felis domestica* or house cat. All the animal properties alluded to in this paragraph and more are being said to apply to a particular animal when we say it is a house cat. This is an enormous condensation, and in that sense simplification, of relevant information. This kind of simplification is characteristic of good classification systems. The absence of a good classification in this sense may mean either that the available information has not been properly organized, or that it is not known, or that the collec- tion of objects to be classified does not permit hierarchical classifi- cation.

In evolutionary theory a hierarchical classification of organisms is used as an indicator of possible genealogical relationships. These classifications are related to distances in multidimensional spaces. There is some freedom in choosing which measurements of each specimen to use and how distance ought to be defined. Differences in these choices have engendered heated argument among taxono- mists (the biologists that are concerned with classifying organisms).[12]

While possible biological misclassifications may not have momen- tous practical importance, simplistic classification into groups may be a serious life or death matter. This is obviously true in medical diagnoses. But there is another important current example. America has recently accepted that there exists a category of persons called "the homeless," and this categorization engenders the corollary "problem of the homeless." At the moment "the homeless" includes a mixture of people defined only by their poverty and misery. The mix includes "addicts" (of many different addictions), chronic drunkards, people that would have been considered clinically insane when there were still insane "asylums," abandoned children, some, times with mothers, and individuals who have lost their homes through unemployment and poverty caused by any of a broad variety of things ranging from bankruptcy to old age. By pooling all these categories, the problems involved gain an impossible complexity, thereby excusing inaction or ineffective action. What has happened is that too few dimensions were used to define proximity of the members of the group. At least a half dozen dimensions were needed, which would have complicated the classification but in exchange would have provided the possibility of relatively more simple solu, tions to some of the problems.

For us, the value of discussing classification is in contributing to the idea of dimensions, which we will need. It does not answer any direct questions about simplicity. Even if a classification is excellent for some purpose, for example, defining evolutionary relationships or making medical diagnosis, it only permits us to initiate the process of ranking in order of simplicity. It by no means assures us that such a ranking is either feasible or useful. Which house cat is more complex than another, and why should we care?

Definition of what are called "complex" numbers relies on the idea of dimension. A brief description will explain the criterion of complexity of numbers, which relates immediately to a classical philosophical definition of complexity.

"Real" numbers can be written as fractions or as strings of inte, gers consisting of numerals plus finite or infinite decimals. Several kinds of "real" numbers can be defined for different uses. There are an infinite number of cardinal numbers (1,2,3, . . .), which are useful for counting objects. There are an infinite number of ordinal numbers for use in ordering a list (first, second, third, . . .). In principle there are an infinite number of what are called "rational numbers," each of which may be precisely expressed by a finite number of decimal places, which are useful for measurement. Notice that there are an infinite number of rational numbers between any two integers (1, 1.1, 1.11, . . .). There are also numbers, called "irrational," that cannot be precisely expressed by any finite number of digits arranged in decimal form. Among these is the ratio of the circumference of a circle to its diameter. In fact there are an infinite number of irrational numbers between any two rational numbers. Which brings us to the terribly curious, but in some sense simple and valid, conclusion that while there are an infinite number of integers, cardinal numbers, rational numbers, and irrational numbers, these infinities are not of the same size.

But what is meant by a "number"? One simplified definition is that numbers are things that give consistent and apparently reasonable results when the arithmetic operations of addition, subtraction, and so on are applied. But if we accept something like this definition, we find that some things satisfy the definition that cannot be written as any arrangement of integers, either fractional or decimal.

As you may recall from junior high school, the result of any number, positive or negative, multiplied by itself must be positive, so there isn't any real square root of any negative number. But multiplying a negative number by minus one makes it positive, changing its sign but not its magnitude. If we can imagine a square root of minus one, then we can express the square root of a negative number as the product of the positive number of equal magnitude multiplied by the square root of minus one. For example, the square root of four is two. The square root of minus four is two times the square root of minus one. The square root of minus four is the same size as the square root of four, but of course it is imaginary. Can we deal with unreal numbers, and why should we want to? There are equations in which the square root of minus one appears in the solution. Fine, then the solution is imaginary, but this does not mean it is not a solution! (Pardon the double negative but it is necessary.) In fact, to deny the legitimacy of these solutions would be to weaken our faith in any solutions, in the sense that we would be saying that mathematical analyses can only be trusted if they agree with our intuitions. But if our intuitions are so trustworthy, do we need analyses?

A reasonably satisfactory solution to the problem is to alter the meaning of the word "imaginary," at least for this context, so that instead of referring to the fantastic and unreal, it has a more mundane meaning. Imaginary numbers, like real numbers, can be ordered along a line. It is a different line from the one on which we order real numbers. Imaginary numbers are on a different dimension from real numbers.

Simple imaginary numbers are then our ordinary positive or negative numbers multiplied by the square root of minus one. Combining imaginary numbers with nonimaginary numbers produces what are called "complex numbers." Complex numbers of this sort cannot be described in terms of only one dimension but require a designation of their position on the two dimensions, the dimension of the real numbers to show their real part and the dimension of the imaginary numbers to show their imaginary part. This is similar in principle to requiring both the dimensions of latitude and of longitude to designate the location of a point on Earth.

Amazingly enough, imaginary numbers turn out to be very useful. Alternating electric current is well described by equations in which complex numbers appear in ways that permit them to be multiplied by each other. Complex numbers are partially imaginary, which is a charming thought by itself, but if the square root of minus one is multiplied by itself, it produces a perfectly real number—minus one—but it slips into being complex again if it once again is multiplied by itself. Imaginary numbers in the right kind of equations twinkle in and out of being real, depending on how they are combined.

What if there are higher orders of complexity of numbers, so that each number is represented by three, four, or five dimensions? What operations still hold? What properties do they have that we have not thought of yet? There obviously are such numbers, but we will not discuss them. ("Curiouser and curiouser!" Suddenly it is no longer surprising that the author of *Alice in Wonderland* was a significant mathematician.)

That takes us far enough to give the idea of one use of "complexity" in mathematics—namely, as a technical term for a particular kind of number system in which several dimensions are involved in each number. However, this is not very satisfying. Complex numbers could have been called something else. If so, how would we know that they were more "complex" in the way I have been using "complex"? Perhaps because they require more symbols to write down than the simple numbers.[13]

This provides a general hint that one way of thinking about complexity in practical contexts is measured in terms of some mechanical property of how things are expressed or written down—

how many symbols, exponents, or equations are needed. Mathematicians and some philosophers (see below) try to define simplicity
in some quantitative way, by counting parts, operations, costs, logical steps, or number of parts into which a thing or idea may be
decomposed.[14] We will return to this in the context of philosophy
in a moment, but there is another recent aspect of mathematical
development that relates immediately to the reality of complexity.

Arithmetic is only one corner of mathematics, however long the
sums. But even humble arithmetic has its share of elegance, when
properly considered. It is easy to understand what it means to add
or subtract, less easy to understand division and multiplication, but
all four operations can be likened to physical activities of a straightforward kind. Fruit is added or subtracted from the greengrocer's
scale. Pies are divided into serving-size portions, and rabbits multi-
ply. This kind of physical understanding of arithmetic is only useful
for the simplest of calculations.

To a large degree, mathematical notation, equations, functions,
and ingenuity are meant to help in making arithmetic calculations.
Try a multiplication problem, say 2,412 by 17, using Roman numerals. Begin by writing the problem as "MMCCCCXII multiplied
by XVII." The innovations of place notation, decimal point, and
the zero as a place holder were tremendous simplifications of the
best sort. One could now immediately distinguish the number two
from the various quantities it might represent (.0002, .2, 2, 200,
2,000,000,000). The use of exponents was introduced just a few
hundred years ago and further simplified life for engineers and
accountants and those mathematicians who deigned to do arithmetic
($2 \times 10^{-4}, 2 \times 10^{-1}, 2, 2 \times 10^2, 2 \times 10^9$). But calculations were
still tedious.

To a large extent intellectually sophisticated mathematics, in so
far as it related to numbers, was historically concerned with finding
equations and the ways to use them (algorithms) that would reduce
the need for arduous calculations. Nevertheless, calculation throughout the ages was enormously consuming of time, money, and human
effort. Even after the development of mechanical calculators, relatively simple problems could still be arduous. This has changed only
in our generation. I recall spending almost a month, using a $500
calculator, inverting matrices that can now be done in seconds using

a hand-held pocket calculator costing less than $30.[15] This massive improvement was made possible by replacing the light-bulb-sized, hot electric valves or gear wheels in archaic computers with transistors, which are tiny and operate on trivial amounts of electric power.

Modern computers take most of the drudgery out of calculation except for truly enormous problems. From the standpoint of at least some applications, the old need for analytic elegance had gone the way of the buggy whip. But, by permitting calculations on a scale previously undreamed of, the idea of elegance has returned in a new form, intimately connected to the idea of complexity. For example, since all computers are finite in their capacities, however big they may be, what are their limits? How can problems be arranged so that the calculation processes (algorithms) are most economical of computer time and expense? What is the relation between the number of entities being considered in a particular context and how many calculations are required for predicting how they will interact? What kinds of problems are likely to exceed the capacities of even the largest of computers? This has generated the new, very active field of "complexity theory," with textbooks, specialists, and the whole apparatus of a new subdiscipline of computing. Further discussions would find us talking *in* rather than *about* these problems, but we at least seem to have found a place where complexity as such is a very practical problem.[16]

In one real but limited sense complexity is measurable in terms of how difficult particular jobs are and also how things are expressed or written down—how many symbols, exponents, equations, and arithmetic operations are needed to perform specific computational tasks. However, this is only obviously valid when performing calculations. Most real-world problems, even including most scientific theories, as contrasted with purely computational problems, concern words, concepts, assumptions of many kinds (not only quantitative ones). These are much less tractable than the problems of mathematicians and are left for philosophers.

Science in general is not as limited as mathematics, but most scientific research is tied to a discrete and finite external subject matter. This subject matter can be drained of further intellectual interest by sufficiently thorough investigation. This does not mean that the subject becomes devoid of value—rather that it seems so well understood that further study is either pointless, unoriginal, or requires more ingenuity than is yet available to establish a sense of novelty. It may then become mummified and wrapped in the sepul-chral prose of textbooks. Parts of Mendelian genetics, ecological population growth theory, classical mechanics, and Euclidian ge-ometry now have or should have this status. Subject matter can also be intellectually unexciting but of possible technological im-portance. Perhaps the gigantic investigation of the human genome will be the sarcophagus for certain kinds of genetics.

This is to be distinguished from an indeterminate kind of schol-arship that grows on itself, developing more subject matter as it goes along.[17] Some philosophers, literary critics, and historiographers (as contrasted with historians) use the works of their fellows as subject matter. Once an assertion has been made, it becomes the subject matter for further discussions, and these in turn are more subject matter for even further discussions. There is the very real possibility of a Malthusian increase in the subject matter for the field if each philosopher writes about previous philosophers and in turn is the subject for future philosophers. Since that is what most philosophers do, in conversations that extend over centuries of thrust and parry usually delivered with wonderful solemnity,[18] I feel even more in-adequate in my discussion of philosophy than in that of science, religion, and historic events. Nevertheless, the discussion must occur because simplicity and complexity are one of the central themes of philosophers.

There are, as you would expect, many different kinds of philos-ophers. Some deal with either the history of philosophy or the philosophy *of* something. So there are philosophers of religion, of science, of ethics, and of technology. (The last I find confusing, but I have several friends who claim to do that, so it is an academic field.) What little training in philosophy I have had was from members of the school whose writings may be almost indistinguish-able from mathematics. These "analytic" philosophers—intellectual descendants of George Boole and Bertrand Russell—are now some-what unfashionable, but they strongly influenced my life when I and they were both in fashion. One of their basic claims was, and is, that the proper concern of philosophy is language—an intellectual struggle to discover what is meant when assertions are made. They often focused on clarification or purification of natural language, or even attempted to construct artificial languages to be used for specific purposes.[19]

The arguably best writings of this school, like the *Tractatus Logico-Philosophicus* of Ludwig Wittgenstein, are minimalist in the best sense, illuminating the process that they themselves exemplify, almost precisely like a painting whose subject matter is the act of painting without being a picture of a painter (see Chapter 5). They may have a poetic beauty but do not lend themselves to reading in bed or on the beach.

But of course all philosophers work with words. Words can be used either in a complex or simple way. Poets' words are rich in associations, so that a line of poetry may be almost impossible to translate from one language to another, and any attempt to paraphrase a poem is much longer than the poem itself. While simplicity is not the primary concern of any philosophical school, it seems to figure most prominently in these two most completely distinct philosophical schools: the analytic and the phenomenological. Analytic philosophers are at the opposite end of the language spectrum from poets. A profusion of meaning jammed into a few words is precisely what they do not want. Their use of words is usually intended to have maximal clarity and precision. They give the effect of dissecting out the entrails of words and spreading them for inspection. Toward the poetic end of the spectrum are the phenomenological philosophers, who will be discussed later.

Analytic philosophers, and more generally those who consider the philosophical analysis of scientific language as their focus, have come to the curious consensus that they do not really understand simplicity in any general way. But their insights into why they cannot develop such an understanding are extremely important, taking us beyond the collection of the anecdotal. Most such studies are reasonably technical, extremely careful, and important for anyone with a deep concern for this area.[20] My summary will focus on general results, without attempting a critical review.

Bunge has made a catalog of different definitions of simplicity as used by philosophers, and one by one finds them to be flawed.[21] For example, complexity cannot be merely quantity in any common sense of the term since, as he says, "Is the water filling a bucket more complex than the water contained in a droplet?" (p. 36). He rejects simplicity measures based on what he terms logical economy, semantic economy of preconceptions, and epistemological economy on the grounds that no numerical function of these makes a "satisfying" measure of complexity (p. 58). He also rejects the opinions of those who use various properties of the written form of a scientific theory as a measure of the theory's complexity. Is the equation for the ellipse twice as complex as that of the hyperbola for having two branches (p. 62)? Is the statement "The world had no beginning" simpler than its contradiction, which requires a theory of beginnings?

Bunge and Sober (and presumably others) deny the importance of the simplicity–complexity distinction as such. One underlying cause of the problem is that even the analytic philosophers are constrained to use natural language for communication and cannot draw the neat boundaries and restrictions of the mathematicians. Theories do not float in isolation. Generally, "there is great difficulty in delimiting what it is that is entailed by any given theory without being strictly deducible from it, since a theory is never precisely formulated in a deductive system, and there are many plausible suggestions and consequences which in a loose sense accompany a theory without being strictly deducible from it."[22]

Adams asserts that "contemporary philosophers of science are convinced that simplicity is a legitimate criterion against which to judge scientific theories, but they are hard pressed to explain why, or even to say what they mean by simplicity" (p. 160).[23] Whatever the virtues of analytic philosophy within the circumscribed boundaries of worlds that resemble the playing field of the mathematicians, when the borders are unclear there seems to be a falling back on intuition to make the decision about what "really" constitutes simplicity and complexity—even from that school of thinkers that is usually so careful to avoid that kind of invitation to polemic. But do the other philosophical schools, less tied to imitation of mathematics, do any better?

Perhaps the most famous philosophical assertions about simplicity and its advantages are those of William of Ockham, in the fourteenth century—the famous "razor of Ockham" or "parsimony principle." This is usually misquoted as "causes shall not be unnecessarily increased" or something similar. Adams's recent study provides me with nice clean quotations which seem to sum up what we need of Ockham and his thought. "There is some puzzle why [this] should be known as Ockham's razor. He did not invent the principle: versions of it are to be found in Aristotle . . . He was not the only medieval to invoke it, and he clearly does not regard it as his principle weapon in the fight against ontological proliferation: rather the Law of Non-Contradiction is" (p. 157).

The following quotations are from translations of work by Ockham (rather than the endless series of restatements that have been made in the past four centuries): "It is futile to do with more what can be done with less." "When a proposition comes out true for things, if two things suffice for its truth, it is superfluous to assume a third." Or "Plurality should not be assumed without necessity," and "No plurality should be assumed unless it can be proved (a) by reason, or (b) by experience, or (c) by some infallible authority." Notice that "authority" meant the Bible, assertions by saints, or rulings by authorized officials of the Catholic Church and that an assertion by these that a cause exists is sufficient evidence as to its existence.

We see that only superficially may Ockham be permitted to pass as a modern philosopher of science (since he also fails to justify his propositions on simplicity). Rather, he was very much a medieval schoolman, carrying an exhausting load of philosophical baggage. For example, he distinguished between cognition of "incomplex" terms such as "bear," "brown," and so-called complex knowledge of them when put together as a proposition, "The bear is brown." One may be cognizant of incomplexa, but knowledge, leading to assent or dissent, is about "complexa" of different levels of complexity. For the medieval schoolmen, knowledge was of several different kinds. Cognition was a kind of knowledge that related to incomplexa, but knowledge of complexa involved volition. But for the scholastics, all of this was embedded in a difficult network of religious dogma and metaphysical assumptions that are deeply intellectual but so far divorced from our everyday thinking as not to be very helpful. Obviously Ockham's sense of what parsimony meant, and for that matter what knowledge itself meant, is different from ours.

I find the idea of "reality" disturbing and inaccessible and feel much more comfortable when raw reality has been pinned down somehow so it lies still to be analyzed. When I say that reality is inaccessible, this does not deny it as a central focus of intellectual concern. Rather, in line with many other examples in this book, it seems more productive to try simplifying it somehow—either by ignoring certain aspects, or subdividing it into more tractable problems—rather than trying to understand it all at once.

While philosophers may accept "reality" as part of their daily work, most would agree that they can say very little about it. Some philosophers discuss it anyway. In our generation these are associated with the school of phenomenologists, with Husserl and Heidegger as the most prominent names. This school reconstructs reality but in a very curious way, quite different from the other simplifications we have encountered. They focus on "being" as such, simplifying the situation in some sense by denying the assertions that stand between the thinker and being. Attributes that can be dealt with scientifically or mathematically are seen as nonessential to "being," and as rapidly as new attributes are suggested they can be discarded as missing the central point of what philosophy "should" be about. This makes for difficult reading.

In fact I have tried to read Husserl and failed ignominiously. First I tried it in college German but rapidly abandoned that goal as pretentious (and also impossible). I tried translations. The words swirled hypnotically and put me to sleep. This happened when I read in the evening, so I decided I had best be fresh and awake— so I tried it in the morning, and once again I was asleep. I tried after coffee and exercise, and then reversed it and exercised first, showered, and then had coffee, but still I fell asleep. I gave up. The successor to Husserl was Heidegger—who seemed a more regular fellow—a strong skier, beloved teacher, and organization joiner as need be. He is somewhat easier to read, but not much, and has generated a mass of explicators and disciples.

I tried them. Some make sense and some do not. Try this quotation from the terminal chapter of an explanatory book on Heidegger by a well-known "humanist" (that is, nonscientist, nonphilosopher).[24] "And when Heidegger intimates a condition of language in which the word was immediate to the truth of things, in which light shone through words instead of being fogged or bent by their dusty use, he echoes exactly Mallarmé's quip made in 1894 (and in fact referred to by Heidegger in one of his late texts) that 'all poetry has gone wrong since the great Homeric deviation.' When Heidegger posits a numinous verity of language in Anaximander, Parmenides and Heraclitus, Mallarmé names Orpheus—of whom, to be sure, no word has survived. The degree to which this 'primalism,' this axiomatic intuition of an earlier stage of authenticity in human affairs—which Heidegger shares with the Marx of the 1848 manuscripts, with the Freud of *Totem and Taboo* and with the Levi-Strauss of *Mythologiques*—represents a secular variant on the scenario of Eden and Adam's fall is an absolutely pivotal question. Its investigation would lead to the root of modern culture."

I do not feel that I am capable of clarifying or summarizing Heidegger with proper depth, but if you are concerned with this kind of thing you should try it for yourself. Unfortunately, as in all the other chapters, this chapter can only point to a great mass of material, like a tour of the Louvre in twenty minutes. The hope is that time will be found to return.

Is it necessarily the case that adding the adjective "really," or even worse the noun "reality," to any sentence is to foreclose further discussion, to declare the game over? Who can stand up against an opponent with reality on his side? In political campaigns, authoritarian religions, peculiarly obscure art criticism, and the assertions of schizophrenics, "reality" figures prominently. The best defense against "reality" used as an intellectual weapon is to hold fast to the realization that legitimate and significant artists, scientists, statesmen, and saints generally point in a direction, but they don't claim to have made the journey.[25]

I must admit to being somewhat disappointed by finding the analytic philosophers falling back on intuition, since, as noted in Chapter 1, intuition seems entwined in self-image, which is tangled in culture. I hoped for a more objective set of standards. Also, the phenomenologists' rejection of most of what seems comprehensible in the name of such a deep reality that very little can be said about it is not incorrect. It is, however, not very helpful.

But perhaps there is another approach—not to philosophy itself but only to the rather constrained problem of thinking about simplicity and complexity. This was suggested by the charming French children's film of the fifties, "The Red Balloon"—which was made into a favorite bedtime book for my children, and now for my grandchildren.[26] The poignant climax involves all of the balloons in Paris rising into the air and going to attend the deathbed and funeral of the hero of the title, who has been struck by slingshot fire. What difference would it make, and in what ways, if all of the balloons in Paris did actually leave for a funeral? Now consider a follow-up film in which the nails and hammers of Paris went to a funeral. It seems clear that the absence of hammers and nails would make a greater difference in more ways to more people than the absence of balloons, regardless of the importance of the balloons to the children.

Is it possible that we could rank objects, people, institutions, species in terms of the complexity of their functions and interactions by considering what would happen if they disappeared? This would perhaps help us out of the world of indeterminate verbiage. But this is just another guess at how we might deal with complexity. At best it will suit a few examples, but certainly not all.

Despite all the problems, it is nevertheless clear that there are such things as complex systems, of differing kinds of complexity. Studies of complex systems may be too narrow when we try to pin them down in terms of standard disciplines and methodologies. The best analyses of complexity are sparkling, exciting, and novel but not necessarily predictive in the usual sense. They may focus on images rather than rigorous theories.

Most rapidly advancing sciences and scholarly fields can be thought of as clearing areas of simplicity in the complex jungle of naive observation. This simplification process often involves focusing on the linear aspects of nature, or at least those that can be considered linear, whence the diversity of processes treated as if they were problems of classical hydrodynamics. Complex systems are generally nonlinear. They can be described but not easily modeled by predictive theories, since differences too small to claim attention can have their effects grossly magnified. The theory of evolution is an example of a nonpredictive but highly descriptive theory for complex systems.

How does complexity of development, structure, and function of organisms emerge from the relative simplicity of biochemistry and genetics? In an essay on the adaptive role of complexity, Huberman, after discussing the generally accepted relation between complexity and hierarchy, presents the intriguing idea that in at least one mathematical sense complexity is an intermediate point between complete order, as of a crystal, and complete disorder, as of a gas.[27]

Various attempts to answer the difficult questions of the origin, control, and evolution of biological complexity involve imaginative borrowings from one discipline to another, attempts to push the limits of nonlinear mathematics to the utmost, and a kind of cou-rageous open-mindedness.

For example, Waddington developed a geometric metaphor for the process of embryological development in which the possible developmental sequences available to an organism are thought of as a series of paths down a mountainous landscape, whose shape is a function of both genetics and environment and in that sense derived from evolution. Sewall Wright at around the same time developed an image of an evolutionary landscape of hills and valleys in which evolutionary fitness increases with altitude, so that all organisms are climbing the hills in evolutionary time, except to the degree that the landscape itself is changing. These two almost poetic images of how complex biological systems change have inspired a rich set of conjectures by biologists, and the mathematician René Thom notes that his own development of mathematical catastrophe theory was inspired by Waddington's epigenetic landscape idea.[28]

In short, complexity and simplicity are real and important and incompletely understood, in many contexts. Approaches to under-standing them may or may not succeed in answering the specific questions being addressed, but I feel that they will almost certainly make important contributions to intellectuality itself.

Closure

Natural scientists share a belief that focusing on simple systems is a good approach to research. On the strength of that belief, for twenty years, on and off, I studied hydra, which the textbooks refer to as the simplest of metazoans.[1] My goal was to develop a formal, analytical, testable mathematical model which predicts aspects of their ecology and future evolution on the basis of a reasonably short list of environmental measurements. The point of the exercise was to demonstrate the possibility of simple theory even in such a complex field as ecology, in contradistinction to simplistic theories or no theory at all. After a sufficient number of false starts and failures, I had some success. The proper next step is to test the theory, but I had begun to wonder if the rather informal assumptions about the intellectually therapeutic effects of simplicity and the definitions of simplicity in organisms were really sound. I had noticed over the decades that often I could not agree with fairly widely held opinions of biologists. Sometimes I could even defend my disbelief. Was simplicity another "vulgar error"?[2] The preceding chapters are notes on my attempt to understand what is meant by simplicity and complexity and what their intellectual role is and has been. I hoped that I could develop a single more or less formal theory of simplicity or, better yet, find one ready-made in, perhaps, the philosophical literature. I found no general theory.

Any attempt to develop a general theory of simplicity and complexity capable of covering the full range of utility of the concepts would resemble Burton's *Anatomy of Melancholy* on one hand and Russell and Whitehead's *Principia Mathematica* on the other, neither of which is meant for easy reading. It would have to begin with a full classification of existing meanings. One would then have

223

to deal with different meanings separately. The result would be encyclopedic, would be enormously technical, and would cost more years than I have left to write. Rather than such a heroic adventure, I found myself on a fascinating, but bewildering, voyage through different intellectual areas, each of which dealt with the ideas associated with simplicity in different ways.

I can summarize what I think I have learned. I will assert fairly strong conclusions without explicit defense, although I believe that their defense is woven throughout the preceding chapters.[3]

All serious attempts to solve discrete problems must involve simplifications. This may mean just focusing attention while ignoring other things, or redefining the problem itself to temporarily obscure its difficult aspects. Simplification may require first developing elaborate intellectual constructs to circumscribe a problem and ensure its tractability. At its best, the attempt to simplify may establish a new viewpoint, from which one may see effective solutions to important problems.

There are, however, dangers in simplification. These range from tripping on a stone because you are reading a book while you walk, to undermining vital national or even global interests because you have chosen an unattainable or useless or selfish but simple goal. It is also possible to simplify problems in a way that will obscure further inquiry.[4] Since simplification is obviously ubiquitous and usually advantageous, among the most important benefits of thinking about simplicity is to be able to learn to discern when simplification is dysfunctional, or at least to be aware of the problem.

On the other hand, "reality," unsimplified and uncomplicated, is generally not accessible except perhaps by some religious insight that is well beyond usual discourse. Sometimes imaginary aspects of "reality" are invoked to simplify issues and problems. Among the favorites are "common sense," "reasonableness," and "human nature." Common sense, not further qualified or tested, is the cause of many social and political problems, or at least has not been very helpful in their solutions. To deny reasonableness as a virtue sounds considerably worse than denying the virtue of motherhood, but until we know more about it in a particular context I see its invocation as a danger signal, roughly analogous to that of entering a restaurant knowing nothing about it other than it claims to have a cook like "Mom."

Appeals to some unspecified theory of human nature must be examined with equal caution, particularly if they are made in a political, nationalistic, or legal context. They often cover up self-interest and various kinds of cultural bigotry with a simplistic concept of human biology and an obfuscated verbal philosophy. Certainly there is no biological evidence for deep common intuitions being part of human nature. The only possible exceptions are agreements which may be based on common physiological or neurological properties.

While the capacity for linguistic communication is contingent on the existence and integrity of specifiable neurological equipment,[5] there is almost nothing automatic about what is said. Even supposedly spontaneous expletives (like saying "Ow" when afflicted with sudden pain) require translation from language to language. Only discourses about truly simple things like hunger, satiety, thirst, warmth, and cold are relatively culture-free, and in that sense intuitive—but even these, almost as they are uttered, are filtered through conventions. I can transmit to almost anyone the idea that I am hungry, but I am likely to find that the food I am offered is inedible, despite my hunger. Only if interactions are brutalized do cultural and linguistic differences lose their importance. Rape does not make the linguistic demands of courtship or even of seduction.

To simplify or complicate involves considered judgment. Groups of people that can share a set of considered opinions, which may be simplifications or complications, constitute "healthy" communities of discourse.[6] Even the best of those teachers who claim to construct ethical, moral, or even merely pragmatic guides for belief and behavior, by appealing to "intuitions" that they claim to share with their hearers, are building on commonality of culture, whether or not they admit it. Even if they are highly educated, but only within a very narrow cultural context, that statement will not be comprehensible.[7] Sufficiently sophisticated teachers will not use an unqualified appeal to intuition in the first place.

Some subjects really are, and some appear to be, more complex than others. Polemics breed in casual obfuscatory simplifications. In apparently complex areas the tendency to oversimplification (that is, simplification that leads to nothing more than nonpredictive equations and meaningless or misleading rhetorical sentences) is very strong, despite empirical evidence. In many fields "discoveries" tend too often to simply be corrections, one at a time, of errors due to oversimplification.[8]

Sometimes, as in the development of the theory of evolution, a brilliant simplifying insight may permit reconstruction of a field. More often the conclusions derived from simplified premises are less valuable, and sometimes not formal enough to be testable at all. To proceed as if the simplification process had not occurred, as if the etiolated entities that emerge from the theory are the same as the full-bodied ones that entered before the simplification, may have serious consequences if the subject matter is at all important.

Unfortunately, it is just the most complex and arguably most important fields of inquiry that are subject to this problem. For example, medicine, ecology, economics, and political science deal with important subject matter, but their very complexity makes gross oversimplification an inviting path.

It might be objected that for some problems there is no other way to proceed. This may or may not be a useful argument. Some-where in Scholom Aleichem's stories, the milkman Tevye appears wearing eyeglass frames. "Why no lenses?" "Can't afford them." "Why wear the frames?" "They're better than nothing." In any case, to not simplify is probably impossible. To not make simplifi-cation completely explicit is obfuscating sleight of hand. To not be sophisticated enough to know it is happening can be dangerous, and to review the examples in this book I hope increases sophistication.

But what have I learned about the handling and care of simplicity, aside from the fact of its importance? In this context, I present an annotated list of what seem to be obvious truths, abstracted from our examples and implicit in our discussion, with some indication of how these may be useful. First are some very general conclusions. Some more immediately practical ones follow.

Complexity and simplicity can be considered in terms of objects, statements, organizations, processes, and ideas, each singly or in any combination. Absolute definitions for complexity and simplicity, without careful specification of context, are not available. Questions such as "Is A simpler than B?" where A and B may refer to statements, equations, religions, or even solutions to mathematical problems may be difficult to answer—and there may be a question as to the value of answers once attained. Specification of contexts and purposes is always valuable and may avoid disasters, but it is not always easy.

Simplification and complication beyond that imposed by neurological mechanisms may be imposed by cultural or artistic convention, for nonpragmatic reasons. Each conventional mode has its own limitations and advantages. For example, linguistic communication (verbal or written) occurs in a linear temporal sequence, which may force inappropriate simplifications on subject matter. Pictures and sculpture escape from linearization, but there is no guarantee that their content can be translated into verbal language.[9] Neither the linear temporal sequence of instrumental music nor the absence of words seems to limit musical complexity in either a formal or communicative sense.

The problems associated with simplicity and complexity are obviously unique in each case, but some generalizations are possible. For example, the approach to simplification differs depending on whether we are concerned with the simplicity of objects or of processes.

To what degree can we extrapolate among different complex systems to aid in their analysis? Organisms are complex systems of physiological, anatomical, and biochemical parts (not mutually exclusive categories). Individual organisms also serve as constituent parts for populations, species, ecological communities (also not neatly demarcated). These biological assemblage systems must be considered as both structures and simultaneously as interwoven nests of processes—exchanging gases, liquids, solids, energy.

There is no overall goal to guide the activities of systems subject to biological evolution, with the obvious exception of humans, who are free to choose their own goals. But the appearance of being designed for a goal is imposed on animals and plants by the selective process as if the goal were persistence in the face of disturbance. Persistence is not constancy.

The only thing that remains constant in organisms is their capacity to persist, and to this constancy all other constancy may be sacrificed. If, and so long as, environmental circumstances do not change their pattern, morphological evolution is very slow. There are many implications of this but for our present purpose—to learn to deal with complex systems—we can examine how they manage this stability. The answer is in the organization of their component parts. Individual organisms all respond to relatively mild disturbances by activating a simultaneous set of responses, some quicker than others. The more rapid responses each generally interferes less with other activities that are being carried on than the slower responses. Problems do not come singly. Organisms may delay considering one problem of great importance while they focus on a more trivial problem of greater immediacy. Immediacy and importance are not necessarily correlated, as the mosquito demonstrates to the working philosopher.

In organisms, if the rapid responses are effective against the disturbance, the slow-starting responses never become fully activated. Does this supply a useful analog or metaphor? It suggests that relative time constants of different parts of complex systems must be examined. It is just those complex systems which show greatest disparity among the time constants of their different parts or subsystems that are in some ways most difficult to analyze, but this is why we must consider them.

For example, with reference to the objects that complicate our society, we are now aware that the life of plastic is too long but the life of paper is too short. Nondegradable discarded plastic is filling garbage dumps while spontaneous decomposition of paper manufactured from acid wood pulp is emptying libraries. It could have been designed differently. This illustrates the need to consider temporal properties of parts of complex systems. Are there any objects or institutions that we would want to be permanent, and what properties must they have? How do we ensure that temporary things really go away?

The causal context of parts of complex systems can also be a potential threat. Much of what is agreed to be ecological disaster has been permitted because the full context of particular technologies was not investigated. The loss of the ozone layer is one of the problems of refrigerator design and hairstyling. Finding alternative refrigerants might have been a serious technical problem, but certainly alternative hair-styling mechanisms might have been developed easily. On the governmental level there is a focus on processes that occur in the time frame of a term of office or a political career. *Aprés mois la déluge,* taken very literally, makes flood-control legislation a low-priority item. To avoid this kind of problem, description of a complex system (whether an object or a process or a procedure that involves objects) should permit answers to specific questions. Specifically, what is its purpose, if it is an artifact? If it is a natural system, what is the motivation for our interest? In what context is the system being considered? Context in this sense also refers to the dimensions in time and space. Is the system supposed to be temporary or permanent? Can it be subdivided into components, and if so are the components more or less complex than the assembled system?[10] What is the sensitivity of the entire system to the properties of its different components? How do the components themselves interact? Is the system part of an even more complex system, and how does that affect our interests?

Our survey seems to suggest that confusion between the simplistic and the simple is a serious public threat. For example, in the United States, the likelihood of simplistically dangerous decisions is materially increased by infatuation with brevity and novelty. American elections are won or lost on the basis of mottoes and thirty-second shots of condensed "information" based on a self-fulfilling assumption made by the opposing parties about the lack of interest by American voters. Both major parties bemoan the fact that the majority of American voters are sufficiently apathetic about this type of treatment that they do not vote at all. No major party acts on the idea that assuming intelligence on the part of the electorate might also be self-fulfilling. This oversight may be self-serving.

Related to our infatuation with the brief shot of information is our emphasis on originality for either its own sake or that of someone's career. The usual penalties associated with outlandish error are considerably less serious than the advantages that come from notoriety.[11] Our love of the outlandish, which permits almost any sufficiently absurd scientific claim, deliberately shocking art, political scandal or allegation of scandal, to fill the papers for weeks may also result from self-fulfilling simplistic assumptions on the part of the media about the intelligence and awareness of the public.

A major source of simplistic solutions to real problems is that the initial problems are displaced by the problem of maximizing the importance of the agencies responsible for their solution. The importance of governmental agencies, universities, scientists, artists and even churches is commonly measured by the simplistic standards of how much money passes through their hands per unit of time.

At a somewhat less simplistic level, persons involved in the different parts of any organization tend to see the primary function of the organization as accomplishing their assigned work. In a university, for example, good janitors measure the quality of different academic departments by how clean they keep their floors, good librarians by how carefully they handle and order books. Good professors tend to be careless about floors. Full discussion of these conclusions would fill at least another volume. Parts of it are contained in texts on operations research and organization theory, but these, as do most textbooks, also tend to simple problems or simplistic reinterpretations of complex ones.

The *New York Times* for January 25, 1990, reported the following main stories about "science": A theory explaining the behavior of child molesters, rapists, and "deviant murderers" (a charming concept in its own right) in terms of a theoretical "lovemap" which is an attribute of everyone but which sometimes develops abnormally; a modification of the position of a group of authors of a 1983 paper that predicted the occurrence of "nuclear winter" in the wake of full-scale atomic warfare, who now assert that it would only be a nuclear "autumn" (which might make some feel better but is bad news for the heavy-sweater industry). In addition, there was a much less exciting, but more considered, story about whether or not Kruger National Wildlife Park is overmanaged. An inside story, which appears to be clear and simple science, reports that lightning seems more frequent and severe in the southern than the northern United States. I am sure that other issues of the *Times* would have a similar set of examples, showing a preponderance of, and more prominence given to, weird and controversial tales compared with those written for more discerning interests. A columnist has compared scientific models with *Vogue* models—"with the right makeup, lighting and layout the ordinary model becomes an irresistibly alluring creature."[12]

Nature tours of all kind are a standard item of entertainment at conferences of professional naturalists. It is as if grand rounds were to be a favorite vacation amusement of physicians. Naturalists assume that a deep, if not necessarily articulate, understanding of nature comes from having observed as many of its manifestations as possible, as if richness of past experience deepens awareness of new experience. From that standpoint the menagerie of examples has been the best reason for me to have written the preceding pages and for you to still be reading. As my journey progressed, I hoped that by examining the concepts of simplicity and complexity from many different paths I could understand them more clearly. Suddenly it is as if I were packing to come home, returning from a trip abroad. The end is almost here.

In current trite terminology, we must reach "closure." Trite terms represent simplifications of real and sometimes important concepts that can be very useful, until we forget what it was they were supposed to be useful about. They then become dangerous, empty reifications or are relegated to menial uses and finally buried in the potter's field of pedantry. For the moment, "closure" is just the term we need. It seems obviously related to the act of closing a box, a suitcase, a discussion, a surgical invasion, and in our case, an open book. Closure transforms something which had been urgent and complex into something that doesn't require ongoing attention at least for a while.

The relation between closure and simplicity is clearly illustrated by the process of packing a suitcase for a journey. The act of beginning to pack is complex by any reasonable definition of complexity. My family and students will assure you that I am not fundamentally neat. My desk, closet, memories, and patterns of association do not normally exist in a clearly organized state. To pack a suitcase before traveling, I must begin with a fair amount of rummaging about, both physical and mental.

Part of the complexity of packing a suitcase resides in its multiplicity of contexts. I must visualize the objectives of my travel—and how the objects in my suitcase might aid in their achievement. But I must first consider the cosmically trivial problems of the mechanics of travel. Can I carry what I am bringing easily or will I need a porter, and do porters still exist on my route? If not, will there be wheeled devices available or should I bring my own wheeled carrier that supposedly folds for carrying on the plane and doesn't when you are balancing your bag on it, but which in my case seems to reverse the process as often as not? Will my luggage meet airlines and customs regulations?

Will I require a proper suit and tie or would a jacket and trouser combination do? Perhaps the jacket would be less wrinkled if I wore it on the plane rather than press it into the bag, but perhaps the hotel will have a valet or perhaps the whole problem is irrelevant because it would be too hot for a jacket in any case? There is a similar train of thought about neckties (never more than two, but never none at all, which says something of the simplicity of professorial life). Socks, underwear, shirts are packed by formula—count the days away and add two. If the sum is less than nine, pack that number; if greater than nine, pack six and assume the existence of laundry facilities. This is all the easy—simple—part.

What is considerably more complex is packing books, manuscripts, data, and charts. This involves an extra layer of mental questions and answers. I will be talking or listening to people; I will be looking at objects or places. What are the questions that will arise? Which of these must be answered on the spot and by material I bring rather than material that I find there?

Finally, how much extra space will I need to bring back gifts for my grandchildren? Can I get that space by jettisoning some of the papers and charts? The operation is dizzying, and I tend to do it in the very early morning when my mind is moderately uncluttered by the problems of today so that I can think of tomorrow's.

When the job is done I reach literal closure. Once locked, suitcases become simple objects, cherished boxes. Their contents are out of immediate consciousness, but I am aware that most of the problems that might arise have been considered. The hectic and complicated mental and physical activity of packing has permitted enormous simplification of the trip itself.

We began with considerations of protohumanity and are concluding with the very modern act of traveling and sightseeing. If we compare modern packing and ancient hand-axe construction as intellectual tasks, we see that packing involves more parts and longer and more tendentious and branching chains of suppositions between my proposed activities and my objectives. As a manual feat, making hand axes is probably more difficult. As an intellectual problem, packing a suitcase is harder. And of course arranging our intellectual luggage is harder still. What is of ultimate importance is distinguishing between the necessary and the customary, between the deeply complex and the only apparently complex, whose complexity can be alleviated by dissecting into simple parts, between devices that really simplify and simplistic solutions that leave us stranded without things we really need. At one level this is best done by manipulating ideas, as if they were pieces on a chess board, but the boundaries of serious playing fields and game boards tend to expand until the players and pieces become very hard to distinguish. Closing the book only contains its problems for a little while.

Lord Darlington. I think life too complex a thing to be settled by these hard and fast rules.

Lady Windermere. If we had 'these hard-and-fast rules' we would find life much more simple.

Lord Darlington. You allow of no exceptions?

Lady Windermere. None!

Lord Darlington. Ah, what a fascinating Puritan you are, Lady Windermere.

Lady Windermere. The adjective was unnecessary, Lord Darlington.

Oscar Wilde, *Lady Windermere's Fan,* Act I, Scene 1

Notes

The Opening

There are those who need a book for the purpose of writing a review, for table conversation, bedroom talk, or whatever but who really do not have enough time, or concern, to read the entire thing. This chapter is for them. It is short enough to complete while standing in a bookstore. The book can be returned to the shelf with a shake of the head, and the reader can pass on to further intellectual adventures and conquests.

1. Benoit B. Mandelbrodt, *Fractals: Form, Chance and Dimension* (San Francisco: Freeman, 1977). Also, an elegant popular survey of this fascinating field is J. Gleick, *Chaos: Making a New Science* (New York: Viking, 1987).
2. J. Aubrey, *Brief Lives,* ed. O. L. Dick (Ann Arbor: University of Michigan Press, 1957).
3. A. Grunbaum, *The Foundations of Psychoanalysis: A Philosophical Critique* (Berkeley: University of California Press, 1984).

1. Sense, Sensibility, and Self

1. "Fitness" in the evolutionary sense will be defined in Chapter 8.
2. J. L. Gould and C. G. Gould, *The Honeybee* (New York: Scientific American Library, 1988), provides a most readable discussion of bee senses and behavior. Parts of the story are, however, controversial, so that A. M. Wenner and P. H. Wells, *Anatomy of a Controversy: The Question of a "Language" among Bees* (New York: Columbia University Press, 1990), should also be consulted. By very careful movement of flowers under one's nose, or, if you have patience, by cutting out the nectar-laden centers and separating them from the petals, you can confirm that there is also a difference in odor in different parts of a flower, which is also more useful if you are the size of a bee than it is for human-sized animals.
3. J. F. Lettvin, H. R. Maturana, W. S. McCulloch, and W. H. Pitts, "What the frog's eye tells the frog's brain," *Proceedings of the Institute of Radio Engineers* 47 (1959): 1940–1951.

4. R. L. Gregory, "Seeing in depth," *Nature* 207 (1965): 16–19. R. L. Gregory, and J. G. Wallace, "Recovery from early blindness: a case study," *Experimental Psychology Society Monograph* No. 2 (1965).

5. P. Heelan, *Space Perception and the Philosophy of Science* (Berkeley: University of California Press, 1983).

6. J. M. Auel, *The Clan of the Cave Bear: A Novel* (New York: Crown, 1980).

7. Alaskan Malamutes are large sled dogs of great charm and small intellect.

8. J. Itani and A. Nishimara, "The study of infrahuman culture in Japan," in *The Fourth International Congress of Primatology*, vol. 1, ed. E. W. Menzel, Jr. (Basel: Karger, 1973).

9. R. A. Hinde and J. Fisher, "Further observations on the opening of milk bottles by birds," *British Birds* 44 (1951): 393–396. Also see J. T. Bonner, *The Evolution of Culture in Animals* (Princeton: Princeton University Press, 1980).

10. Beaver dams and prairie-dog tunnels do persist, but individual and persistent innovations in their design have not been demonstrated.

11. M. Merleau-Ponty, *The Primacy of Perception and Other Essays*, ed. J. M. Edie (Evanston: Northwestern University Press, 1964), pp. 126–127.

12. I have discussed the role of self-awareness and normative introspective self-image elsewhere. See "The peculiar evolutionary strategy of man," *Boston Studies in Philosophy of Science*, 71 (1983): 227–248. "Evolution is no help," *World Archaeology* 8 (1977): 333–343. "Is history a consequence of evolution?" *Perspectives in Ethology* 3 (1978): 233–255. "Species identity, self image, and the origin of persons," in *Species Identity and Attachment: A Phylogenetic Evaluation*, ed. M. Roy (New York: Garland STPM Press, 1979).

13. G. Gallup, "Self recognition in primates," *Am. Psychol.* 32 (1977): 329–338.

14. There are serious moral issues involved in caging chimpanzees, but that is another problem.

15. R. Epstein, R. P. Lanza, and B. F. Skinner, "Self-awareness in the pigeon (*Columbia livia domestica*)," *Science* 212 (1981): 695–696, demonstrated that under elaborate circumstances pigeons could be operant-conditioned to refer to their own appearance. I do not believe that behavior elicited under operant conditioning is of serious relevance to field conditions, any more than the playing of horns by circus sea lions is relevant to either music or marine biology.

16. J. Goodall, *The Chimpanzees of Gombe* (Cambridge: Harvard Univer-

sity Press, 1986), p. 123. L. Godfrey, "Evolution and ethics—an ape's eye view," in *Evolution and Ethics,* ed. M. Nitecki (Chicago: University of Chicago Press, in press), and other papers in the same volume.

17. The pre-necrotic Bronx is somewhat idealistically described in L. Ultan, *The Beautiful Bronx* (New York: Crown, Harmony Books, 1979).

18. Goodall, *Chimpanzees of Gombe.*

19. Summaries of dating methods are available in any of the many popular books on human evolution. These include J. Reader, *Missing Links* (New York: Little, Brown, 1981), which is a beautifully illustrated popular history of the hunt for fossil hominids, and R. Lewin, *Bones of Contention* (New York: Simon and Schuster, 1987), an account of very recent finds and arguments.

20. P. R. Jones, "Experimental implement manufacture and use: a case study from Olduvai Gorge, Tanzania," *Phil. Trans. Roy Soc. London B* 292 (1981): 189–195.

21. Of course the stone tools manufactured by humans, like the Danish dagger, were made with very high expertise, first-quality stone, and a great deal of time. All the standards of craftsmanship are relevant. This discussion is not about human craftsmen but rather about the beginnings of the capacity for craft.

22. This has an obvious weakness. J. B. S. Haldane has written somewhere that if we really want to infer the nature of the Creator from His creation, we must believe He had an inordinate fondness for beetles!

23. T. G. Wynn, *The Evolution of Spatial Competence,* Illinois Studies in Anthropology, No. 17 (Chicago: University of Chicago Press, 1989).

24. Wynn, *Evolution,* p. 99. Notice that he does not claim that the 300,000-year-old tools were actually made by what we would consider to be legally human beings, perhaps because the most recent tool sets that he analyzed were not associated with human fossils.

25. K. P. Oakley, "Emergence of higher thought 3.0–0.2 Ma B.P.," *Proc. Roy Soc. London B* 292 (1981): 205–211, and R. A. Dart, "The waterworn australopithecine pebble of many faces from Makapansgat, *South African Journal of Science* 70 (1974): 167–169.

26. An excellent recent summary of fossil evidence and evidence from modern nervous systems and how these relate to the origin of human intellectual capacity is John C. Eccles, *Evolution of the Brain: Creation of the Self* (London: Routledge, 1989). This is a serious work by a major intellectual which can be read by a lay reader.

27. Theatrical miming demonstrates that narratives can be transmitted without words, but audiences at mime performances carefully read their program notes.

28. R. S. Solecki, *Shanidar—The First Flower People* (New York: Knopf, 1971). Other early burials are known, and burial is not the only way of demonstrating humanity. Archaic humans demonstrated their humanity at various places and in various ways during the interval from around 100,000 to 30,000 years ago. By the latter date all humans were intellectually essentially indistinguishable from moderns, as far as can be determined from paleontological or archaeological evidence. The transition from late archaic to early modern humans, including social, artistic, physical, and ecological changes, is a vibrant and rapidly changing research area. An excellent recent summary can be found in E. Trinkaus, ed., *The Emergence of Modern Humans* (New York: Cambridge University Press, 1989). J. M. Auel, the novelist (see note 6), was a supporter and participant in the conference which this book summarizes.

29. C. Taylor, *Sources of the Self: The Making of Modern Identity* (Cambridge: Harvard University Press, 1989). I will return to further discussion of this study in the last two chapters.

30. R. Alexander, *Darwinism and Human Affairs* (Seattle: University of Washington Press, 1979). Also, C. J. Lumsden and E. O. Wilson, *Promethean Fire: Reflections on the Origin of Mind* (Cambridge: Harvard University Press, 1981), is the most relevant of several possible references since it summarizes more technical earlier arguments for the general reader. I reviewed this book in *Evolution* 38:461–462. I have commented on some of Alexander's arguments in L. B. Slobodkin, "Problems on the border between biological and social sciences," Michigan Discussions in Anthropology 2 (1977): 124–137.

31. L. B. Slobodkin, "The complex of questions relating evolution to ethics," in *Evolution and Ethics,* ed. M. Nitecki (Chicago: University of Chicago Press, in press). See also note 12, above.

2. Simplifying Religious Revolutions

1. Doctrines usually are surrounded by artistically and emotionally satisfying rituals, which will be discussed in course. Here I am acting as if all that mattered was doctrine.

2. Orwell saw the rewriting of history as a major evil, and having it done by amateurs may be even worse. But ongoing reappraisal of history is not necessarily part of a cynical political plot. Historical accounts of events which occurred in the context of an archaic deep belief system must be written again and again as deep beliefs change, if the events are to be at all comprehensible. This is why there is no definitive history. I have attempted to understand and to respect the deep belief

systems that may have underlain the doctrinal histories. It is only if we make that attempt that the historical changes in religious doctrine can be considered as we would consider changes in other intellectual constructs.

3. J. C. Carrol, *Slave Insurrections in the United States, 1800–1865* (New York: Negro Universities Press, 1938), gives accounts of many slave insurrections. Some of the leaders, like General Gabriel and Nat Turner, were either privileged or at least had distinguished themselves in some way prior to the insurrections, but further information as to whether or not a relatively privileged position of the leaders was a statistically significant property of slave insurrections would be of interest. P. Levi, *The Drowned and the Saved* (New York: Summit Books, Simon and Schuster, 1988), trans. R. Rosenthal, does assert that the special squads were more likely to rebel. Crane Briton, *The Anatomy of Revolution: Revised and Expanded Edition* (New York: Prentice-Hall, 1952).

4. An elegant series of essays on this general subject were published as *Wisdom, Revelation, and Doubt: Perspectives on the First Millenium B.C.* in *Daedalus: Journal of the American Academy of Arts and Sciences,* vol. 104, no. 2, 1975. For the rest of this chapter I will repeatedly refer to specific papers in this volume. The novel is Gore Vidal, *Creation,* (New York: Random House, 1981).

5. V. Nikiprowetzky, "Ethical Monotheism," p. 89 (see note 4).

6. Of course the pursuit of originality can also become a doctrinal orthodoxy, as discussed in Chapter 6.

7. W. R. Cross, *The Burned Over District: The Social and Intellectual History of Enthusiastic Religion in Western New York, 1800–1850* (New York: Farrar, Straus & Giroux, 1950).

8. E. Pagels, *The Gnostic Gospels* (New York: Vintage Press, Random House, 1981).

9. I see even such obviously Jewish theologians as Martin Buber and Yehoshua Heschel as writing in response to non-Jewish theological and philosophical concerns. This in no ways diminishes their global importance. I am merely asserting that mainstream Judaism is legalistic and Pharisaic. There are mystical schools and texts and there always have been, but these do not connect neatly with Western theology.

10. Both the content and intellectual history of the Kabbalah are enormously complex. My brief comments should not be treated as even a summary or definition. For those concerned with this subject, I recommend the recent book by Moshe Idel, *Kabbalah: New Perspectives* (New Haven: Yale University Press, 1988).

11. Isaac Luria's influence was a particular focus of Gershom Scholem,

whose works are probably the main source of information about kabbalah for Anglophones, for example, *Kabbalah* (New York: Quadran-
gle, 1974). Idel puts this work into a broader perspective (see note 10).

12. The development of Christianity and rabbinic Judaism occurred at
around the same time. The rabbis were continuing their reconstructing
of a priestly religion into a legalistic one while the Church fathers
were separating themselves from Judaism. It was not a matter of
Christianity's emerging from full-fledged rabbinism.

13. W. H. C. Frend, *Saints and Sinners in the Early Church: Differing
and Conflicting Traditions in the First Six Centuries* (Wilmington:
Michael Glazier, 1985), is a fascinating account of early Christian
doctrinal history.

14. Some aspects of what we now see as typically Islamic artistic style
were also in place before the Koran was codified, as in the Mosque of
the Rock. The prohibition of representational art and the enormous
complication of calligraphic art was considerably later.

15. Note that this is anachronistic. Samaritan and rabbinic Jews separated
only after the Babylonian exile at the earliest.

16. L. Dumont, "Understanding non-modern Civilizations," pp. 153–172
(see note 4), considers that the role of "renouncer," a monastic individual outside the caste system, was a fairly general one in ancient
India. Jaina and Buddha are particularly interesting examples.

17. P. S. Jaini, *The Jaina Path of Purification* (Berkeley: University of
California Press, 1979), is a scholarly discussion of Jainism by an
insider. J. Campbell, *The Masks of God: Oriental Mythology* (1962; rpt.
New York: Penguin, 1982), is an encyclopedic study of oriental religions. Campbell is a slightly unsympathetic outsider to Jainism.

18. Patricia Crone and Michael Cook, *Hagarism: The Making of the Islamic
World* (New York: Cambridge University Press, 1977).

19. Ahad Ha-'Am (Asher Ginzberg), "Sacred and profane," in L. Simon,
trans., *Selected Essays by Ahad Ha-'Am* (Philadelphia: Jewish Publication Society of America, 1944).

20. Cited by P. Brown, "Society and the supernatural," p. 146 (see note
4) from Vogel, *Le pecheur et la penitance au moyen age* (Paris, 1969).

3. The Great Intellectual Playing Field

1. There is a supportable belief among some historians that the political
manipulative skills of acknowledged geniuses may be as important as
their intellectual depth. See D. L. Hull, *Science as a Process: An*

Evolutionary Account of the Social and Conceptual Development of Science (Chicago: University of Chicago Press, 1988). Nevertheless, intense mental ability and effort still have a role.

2. These are available in English: *Waning of the Middle Ages: A Study of the Forms of Life, Thought and Art in France and the Netherlands in the XIVth and XVth Centuries,* trans. F. Hopman (London: E. Arnold and Co., 1924), and *Homo Ludens: A Study of the Play Element in Culture* (Boston: Beacon Press, 1955).

3. W. C. Dilger, "Evolution in the African parrot genus *Agapornis,*" *Living Bird* 3 (1964): 135–148. The genera *Loriculus* and *Agapornis* carry nesting materials tucked in their feathers. *Loriculus* sleeps upside down, preventing the bits of nesting material from falling out.

4. C. Elton, *Animal Ecology* (London: Sidgwick and Jackson, 1927).

5. *Tom Sawyer* (New York: Harper and Brothers, 1922), p. 15.

6. Recall Chapter 1.

7. M. Rosenzweig reminded me that the Abbott and Costello 1941 film "Buck Privates" contains a sequence of a poker game in which Abbott keeps changing the rules with fine comic effect.

8. Actual games are serious enough so that their corruption is not funny, except to spectators or in fiction. The game of croquet in *Alice through the Looking Glass* in which the balls are hedgehogs and the mallets are flamingos that act of their own volition angered Alice.

9. A. Rapoport, *Fights, Games and Debates* (Ann Arbor: University of Michigan Press, 1960), is particularly good in clearly distinguishing the role and limitations of game theory.

10. Obviously, rules can be changed, but this is done between, not during, games and must be known to all players before the next game starts.

11. Yasunari Kawabata, *The Master of Go* (New York: Perigee Books, 1981), and V. Nabokov, *King, Queen, Knave* (New York: McGraw Hill, 1968).

12. I know that football and mountain climbing and hang gliding are physically less safe than lying in bed or sitting at a desk, but not all dangers are physical.

4. Three Dinner Parties

1. See Chapter 9, in which I discuss the relation between virtue and simplicity and vice and complexity.

2. Tatsuo Motokawa, "Sushi science and hamburger science," *Perspectives in Biology and Medicine* 32 (1989): 489–504.

3. The terminal bit of matzoh is called by the Greek word "Aphikomen," which is traditionally understood to refer to the rather lively parties that Greek youth would attend after a symposium meal. The intent is that the evening's entertainment for the Jews is all at the seder table— an enforcement of simplicity in amusements.

5. A Matter of Taste

1. V. Newhouse, *Wallace K. Harrison: Architect* (New York: Rizzoli, 1989).

2. A. K. Coomaraswamy, *The Christian and Oriental or True Philosophy of Art: A Lecture Given at Boston College, in March 1939* (Newport, Rhode Island: John Stevens, 1939), p. 5.

3. For example, the beautiful and gigantic six-pound volume, *History of Modern Art,* by H. H. Arnason (New York: Harry N. Adams, Inc., 1977), confines its attention to the nineteenth and twentieth centuries, Europe and America, only, and still omits many artists.

4. Compare the discussion of indeterminately growing intellectual fields in Chapter 10.

5. L. Steinberg, "Art and science—do they need to be yoked?" in *Daedalus,* summer 1986, pp. 2–16.

6. Mark Twain, *A Tramp Abroad,* vol. 2, chapter 21, "Titian Bad and Titian Good." (New York: Harper & Brothers, 1899).

7. Arnason, *History,* p. 64.

8. Erich Auerbach, *Mimesis: The Representation of Reality in Western Literature,* trans. William Trask (Princeton: Princeton University Press, 1953).

9. Oscar Wilde's *Portrait of Dorian Gray* focuses on the relation between picture, person, and morality.

10. E. T. Gilliard, *Birds of Paradise and Bower Birds* (Garden City: Natural History Press, 1969).

11. See Chapter 3, note 2.

12. Pp. 229–231 in the Time-Life Books edition, 1964.

13. Pseudo-Lucian, "Affairs of the Heart," in *Lucian,* vol. 8, trans. M. D. Macleod, Loeb Classical Library (Cambridge: Harvard University Press, 1974).

14. The distinction between critic, philosopher of art, and art historian is not always clear. A recent work that I found particularly lucid and of singularly low pomposity is by Lucian Krukowski, *Art and Concept: A Philosophical Study* (Amherst: University of Massachusetts Press, 1987). I was relieved to find that the definition I present for art could

have been derived from that work or from the essays by Adrian Stokes, *The Image in Form: Selected writings,* ed. R. Wollheim (New York: Harper and Row, 1972).

15. This is widely pictured in tourist folders and the like. It is also on p. 57 in T. Sawa, *Art in Japanese Esoteric Buddhism* (New York and Tokyo: Weatherhill/Heibonsha, 1972).

16. See M. Critchley and R. A. Henson, eds., *Music and the Brain: Studies in the Neurology of Music* (London: William Heinemann Medical Books, 1977), examines this problem in some detail. I found most relevant the introductory chapter by Henson, pp. 3–21, and Chapter 12, "Music, emotion, and autonomic function" by G. and H. Harrer, pp. 202–216.

17. A recent collection of his drawings along with photographs of some of his mobile sculptures can be found in Rowland Emett, *From Punch to 'Chitty-Chitty-Bang-Bang' and Beyond* (London: Chris Beetles Ltd., 1988).

18. Theatre has many more layers between the author and the performance.

19. There are many books and more coming out all the time. Krukowski (note 14) is very good. John Cage's writings and people writing about him are also lucid and original. Also R. Kostelanetz, ed., *John Cage,* Documentary Monographs in Modern Art (New York: Praeger, 1970), and K. Baker, *Minimalism, Art of Circumstance* (New York: Abbeville Press, 1988), is a great gallery of works.

20. C. Brinton, *The Anatomy of Revolution: Revised and Expanded Edition* (New York: Prentice-Hall, 1952), provides an analysis of the French, Russian, American, and British political and military revolutions, but his analysis seems suggestive for revolutionary movements in general.

21. C. Griswold, "The Vietnam Veterans Monument on the Washington Mall: Philosophical thoughts on political iconography," *Critical Inquiry,* 12 (1986): 689–719, is an elegant essay on the meaning and role of the Memorial.

6. *To Science Is Human*

1. J. Needham, *Background to Modern Science* (Cambridge: Cambridge University Press, 1938). G. Sarton, *Sarton on the History of Science: Essays Selected by Dorothy Stimson* (Cambridge: Harvard University Press, 1962).

2. J. Watson *The Double Helix: A Personal Account of the Discovery of the Structure of DNA* (London: Weidenfeld and Nicholson, 1968). F. Crick, *What Mad Pursuit: A Personal View of Scientific Discovery* (New York: Basic Books, 1988).

3. *The Ngor Collection of Ngo Thar rise mKhan po bSod nams rgya mtsho* (Tokyo: Kodansha Ltd., 1983) is the most beautiful collection that I have seen in print.

4. T. S. Eliot, *Four Quartets* (London: Faber & Faber, 1979), p. 26.

5. D. Eisikowitch, "The role of dark flowers in the pollination of certain umbelliferae," *Journal of Natural History* 14 (1980): 737–742.

6. A. C. Crombie, *Medieval and Early Modern Science* (Garden City: Doubleday, 1959), pp. 186–187.

7. Louis Slobodkin, *Sculpture: Principles and Practice* (New York: Dover Press, 1973).

8. These issues are of particular concern to historians of science with a Marxist interest.

9. P. Martin and R. G. Klein, eds., *Quaternary Extinctions: A Prehistoric Revolution.* (Tucson: University of Arizona Press, 1984).

10. Joseph Needham and his colleagues have produced a prodigous volume of material on and about Chinese scientific history. Starting in 1954, volumes began appearing summarizing these studies: J. Needham, *Science and Civilization in China* (Cambridge: Cambridge University Press, 1954). A popularization and summary of this material can be found in R. K. Temple, *The Genius of China: 3000 Years of Science, Discovery, and Invention* (New York: Simon and Schuster, 1986).

11. E. Topsell, *The History of Four Footed Beasts, and Serpents, and Insects* (New York: Da Capo Press, 1967), is a late English example of combined natural history and homiletics. Also see J. Gerard, *The Herbal or General History of Plants: The Complete 1633 Edition as Revised and Enlarged by Thomas Johnson* (New York: Dover Press, 1975).

12. The lipstick case had a small hole in the bottom and a cork at the other end so that it floated upright, air-filled and almost submerged. When the rubber dam was in place over the medicine bottle mouth, pressure on the dam increased air pressure over the water, which in turn forced a little water into the bottom hole of the lipstick case, constricting the enclosed air, and slightly increasing the weight per unit mass of the diver, which then sank. By very delicate adjustments of pressure on the rubber we could make the diver almost stay at a fixed depth. This beautifully demonstrated ideas of buoyancy, density, and the mystery of floating in swimming pools.

13. W. A. Greenhill, ed., *Sir Thomas Browne's Religio Medici, Letters to a Friend &c. and Christian Morals* (London: Macmillan, 1950), p. 24.

14. W. Harvey, *Anatomical Studies of the Motion of the Heart and Blood,* trans. Chauncey D. Leake (Springfield: Charles Thomas, 1949), p. 46.

15. Richard Hale, "The Humane Allantois fully discovered . . . With an answer to their objections, who deny it still," *Philosophical Transactions, Royal Society of London,* no. 271 (1701): 835–850

16. Anton van Leeuwenhoek, *Alle de Brieveen van Anton van Leeuwenhoek* (Amsterdam: Smets and Zeitklinger, 1941), vol. 2, pp. 291–293.

17. The rapid growth of ecology has been documented in R. B. Root, "The challenge of increasing information and specialization," *Bull. Ecol. Soc. America* 68 (1987): 538–543.

18. The use of the word "theory" in science will be made understandable in the next two chapters.

7. Explaining the Whole Universe

1. Gershom Scholem, *Kabbalah,* and Moshe Idel, *Kabbalah: New Perspectives* (see notes 10 and 11, Chapter 2).

2. Three or four Nobel Laureates, a half dozen MacArthur Fellows, and a handful of eminent authors, artists, and musicians.

3. A. G. Gross, *The Rhetoric of Science* (Cambridge: Harvard University Press, 1990), pp. 168–174.

4. T. Kuhn, *The Structure of Scientific Revolutions* (Chicago: University of Chicago Press, 1962).

5. L. Slobodkin, "Natural philosophy rampant: An essay review of *Ecological Communities: Conceptual Issues and the Evidence.* edited by D. R. Strong et al.," *Paeleobiology,* 12 (1986): 112–119.

6. L. Slobodkin, "Listening to a symposium: A summary and prospectus," in M. H. Nitecki, ed., *Biotic Crises in Ecological and Evolutionary Time* (New York: Academic Press, 1981), pp. 269–288.

7. Galileo Galilei, *"On Motion and On Mechanics"; Comprising "De Motu" (ca. 1590) and "Le Meccaniche" (ca. 1600),"* trans. with Introduction and Notes by I. E. Drabkin and Stillman Drake (Madison: University of Wisconsin Press, 1960), p. 29.

8. Morris Kline, *Mathematics: The Loss of Certainty* (New York: Oxford University Press, 1980).

9. Alexander Pope, "Epitaph Intended for Sir Isaac Newton," in *Minor Poems of Alexander Pope,* ed. Norman Ault, completed by J. Butt (New Haven: Yale University Press, 1970), p. 317.

8. Explaining the Rest of the Universe

1. For fuller discussion of this subject, and the more technical details of evolutionary theory, please refer to either a standard text, like Douglas Futuyma, *Evolutionary Biology* (Sunderland: Sinauer, 1986), or a popularization like Richard Dawkins, *The Blind Watchmaker* (London: Penguin, 1986). There is an abundant, and sometimes cranky, polemical literature which can confuse novice readers dreadfully; these two references were chosen with care.

2. The questions of why some very simple-seeming animals can manage as well, in an evolutionary sense, as much more complicated ones is where this book began, since my empirical research for some years has focused on the polyps of brown and green hydra. This book summarizes the problems I encountered when I thought more deeply about what was meant by "simple."

3. Robert Hooke, *Micrographia* (Weinheim: J. Cramer, 1961; facsimile of 1665 edition), Swammerdam, *Historia Insectorum Generalis* (Leyden: Apud Isaac Severinum, 1693), and most of the issues of the *Philosophical Transactions of the Royal Society of London,* prior to around 1720, are very much worth looking at. A particular pleasure comes from the fact that many major university and museum libraries have good facsimile editions so that turning the pages now permits one to almost feel the excitement they must have generated when they were new.

4. J. Bonner, *The Evolution of Complexity by Means of Natural Selection* (Princeton: Princeton University Press, 1988), is the most recent serious study of the complexity of organisms in an evolutionary context. References to Bonner, throughout this chapter, refer to this work.

5. I have discussed "evolutionary success" in earlier papers. For example: L. B. Slobodkin, "The strategy of evolution," *American Scientist,* 52 (1964): 342–353; L. B. Slobodkin, "Toward a predictive theory of evolution," in R. Lewontin, ed., *Population Biology and Evolution* (Syracuse: Syracuse University Press, 1968); L. B. Slobodkin and Anatol Rapoport, "An optimal strategy of evolution," *Quarterly Review of Biology* 49 (1974): 181–200. The earliest serious analysis of the inverse relation between longevity and possible rates of population increase that I know of is by F. E. Smith, "Quantitative aspects of population growth," in E. Boell, ed., *Dynamics of Growth Processes* (Princeton: Princeton University Press, 1954). While the inverse relation between longevity and reproductive rate is certainly real, it is by no means obvious why it should be so.

6. N. H. Sleep, K. I. Zahnie, J. F. Kasting, and H. Morowitz, "Annihi-

lation of ecosystems by large asteroid impacts on the early earth," *Nature* 342 (1989): 139–142. They estimate that any possible life was destroyed by the effects of the impact of large asteroids on earth between 4.0 and 3.8 billion years ago and that current biological history began around 3.5 billion years ago, give or take a half billion years. A half billion years is almost surely enough for the spontaneous origin of life on a lifeless planet to have a very high probability, without re-quiring anything outlandish in the way of environment.

7. Richard Dawkins, *The Blind Watchmaker* (London: Penguin, 1986), p. 123.

8. The early history of the theory of evolution is fascinating. The first publication of modern evolutionary theory was C. Darwin and A. R. Wallace, "On the tendency of species to form varieties; and on the perpetuations of varieties and species by natural means of selection." Communicated by Sir Charles Lyell and J. D. Hooker; *Journal of the Linnean Society, Zoology* 3 (1859): 45–62. It is very much worth reading. It includes a letter Wallace sent Darwin in which the theory of evolution is expounded in much the same way as in this chapter, two testimonials from Darwin's close friends attesting to Darwin's character and to the fact that he had been working on this subject for some time prior to receiving Wallace's letter, and two excerpts of manuscripts by Darwin. J. L. Brooks, *Just before the Origin: Alfred Russell Wallace's Theory of Evolution* (New York: Columbia University Press, 1984), presents strong support for the importance of the role of Wallace. E. Mayr, *The Growth of Biological Thought* (Cambridge: Harvard University Press, 1982), presents the more usual view. A lively popular account is provided in S. J. Gould, *Ever Since Darwin* (New York: Norton, 1977).

9. My children have gotten approximately half their genetic material from me and a trifle more from my wife. Most probably around one fourth of the genetic material that my grandchildren have came from me, but there is an element of chance in this. Each grandchild may actually be carrying copies of anywhere from all to none of the genes that I contributed to its parent.

10. The wording of this paragraph was improved by M. Rosenzweig.

11. I do not want to discuss elementary genetics at any length, but further material on individual genetic differences due to recombination can be found in any genetics text. In brief, since the members of a pair of diploid chromosomes enter gametes at random, and since there are several pairs of chromosomes in each parent, the number of possible different genotypes that may arise from mating two genetically distinct

parents is enormous. Assuming no linkage between genes, a monohybrid cross results in three possible genotypes, a dihybrid cross in nine, and in general a cross between parents, each hybrid for n genes, can produce 3^n genotypically different offspring. Given twenty-three pairs of chromosomes in humans, and making the reasonable assumption that the two parents are hybrid for at least one set of genes on each chromosome pair, then the number of genotypically possible organisms these parents can produce is equal to or greater than 3^{23}, which is a very large number, considerably larger than the number of people that have ever lived.

12. Barbara McClintock, "Chromosome organization and genetic expression," *Cold Spring Harbor Symposium on Quantitative Biology* 16 (1951): 13–47; and "Controlling elements and the gene," *Cold Spring Harbor Symposium on Quantitative Biology* 21 (1957): 197–216; and "The control of gene action in maize," *Brookhaven Symposium in Biology* 18 (1965): 162–184. McClintock showed that particular genes in corn can cause other genes to mutate to a limited number of alternative states. She was awarded a Nobel Prize and a MacArthur Fellowship (significantly in that temporal order) more than thirty years after the start of this work. These and similar genes, which are now being studied intensively, can also shift their locations from chromosome to chromosome. This exciting story is still unfolding.

13. I recently found that I can successfully run a washing machine from a set of directions written in the Japanese secondary schools' version of English. (But in the first draft of that sentence I typed "ruin" in place of "run," demonstrating that this, like most analogies, must be handled gingerly.)

14. A good introduction to the problem of assembling unlikely looking properties by nonteleological mechanisms is in W. C. Wimsatt and J. Schank, "Two constraints on the evolution of complex adaptations and the means for their avoidance," in M. Nitecki, ed., *Evolutionary Progress* (Chicago: University of Chicago Press, 1988), pp. 231–273. Also, sea lions, not baboons, play tunes on horns at circuses. The use of baboons in the limerick and my discussion of it was poetic license. Sea lions are stupid enough to be trained by the crude pedagogy used in circuses, while baboons may become dangerous and resentful.

15. The unfortunate "sociobiology" controversies of the 1970s were mainly based on E. O. Wilson, *Sociobiology: The New Synthesis* (Cambridge: Harvard University Press, 1975), and the responses to it. It was difficult to avoid becoming somewhat involved. As noted in Chapter 1, I do not believe that there is any proper genetic evidence for the validity of Wilson's theory.

16. Even severe-seeming environmental changes, if they have occurred with sufficient regularity and a relatively high frequency, may not have selective effects. See L. Slobodkin and A. Rapoport, "An optimal strategy of evolution." *Quarterly Review of Biology* 49 (1974): 181–200.

17. As in Futuyma's text (see note 1, above).

18. D. Futuyma, "On the role of species in anagenesis," *American Naturalist,* 130 (1987): 465–473; G. E. Hutchinson, "When are species necessary?" in R. Lewontin, ed., *Population Biology and Evolution* (Syracuse: Syracuse University Press, 1968), pp. 177–186.

19. The mechanisms of evolution as listed are the properties of life itself. Clearly then the origin of life from the nonliving involves other mechanisms. Research on these mechanisms is going forward, but there is as yet no consensus. See H. Morowitz, *Mayonnaise and the Origin of Life* (New York: Scribner, 1985).

20. This relation between landscape color and English sparrow color is sufficiently obvious that Nelson Hairston, Sr., Fred Smith, and I borrowed skins from the University of Michigan Museum and let a freshman biology class discover it for themselves.

21. H. Carson, "Chromosomal sequences and inter-island colonizations in Hawaiian Drosophila," *Genetics* 103 (1983): 465–482, provides the evidence for speciation occurring on the islands. There is an ambiguity about what "time for speciation" means. Here I mean the time from the splitting off of a species to the time that a second species splits off from it. The other conceivable meaning—the time between the start and the end of the splitting process—carries the implication that, once started, the process of splitting is inevitably going to result in an actual speciation. This is not so.

22. A. Meyer, T. D. Kocher, P. Basasibwaki, and A. Wilson, "Monophyletic origin of Lake Victoria cichlid fishes suggested by mitochondrial DNA sequences," *Nature* 347 (1990): 550–553, and personal communication with Meyer.

23. See almost any evolution text. There is an amazing series of difficult-to-read volumes by Sewall Wright himself published during the interval 1968–1978 entitled *Evolution and the Genetics of Populations: A Treatise* (Chicago: University of Chicago Press). Also a classic early work on the mathematical theory of evolution has just been reprinted: J. B. S. Haldane, *The Causes of Evolution* (Princeton: Princeton University Press, 1990).

24. My hope in using hydra was to dodge these problems. Partial fruition of that hope is presented in M. Gatto, C. Matessi, and L. B. Slobodkin, "Physiological profiles and demographic rates in relation to food quan-

tity and predictability: An optimization approach," *Evolutionary Ecology* 3 (1989): 1–30.

25. W. B. Provine, *Sewall Wright and Evolutionary Biology* (Chicago: University of Chicago Press, 1986).

26. J. G. Frazer, *The Golden Bough* (New York: Macmillan, 1922); J. Campbell, *The Masks of God* (New York: Viking Press, 1959); Herodotus, *The History of Herodotus* (Chicago: Encyclopedia Britannica Press, 1952).

27. The lovely Biblical nineteenth psalm can be read as an invitation to scientific inquiry and an acknowledgement of the role of the human intellect, very similar to the passage from Browne quoted above. In the Hebrew the second verse carries the obvious sense that the hosts of heaven, the stars, declare the glory of God, but they have neither voice nor hearing nor do they make a sound except through the intellect of humanity. This image apparently made no sense to the translators, who added extra words so that the English now reads as if the stars sing like a hybrid between the Mormon Tabernacle Choir and the twittering birds in a Walt Disney movie.

28. C. Lyell, *Principles of Geology,* 11th ed. (New York: Appleton, 1872); J. Hutton, *Theory of the Earth with Proof and Illustrations* (London: Cadell and Davies, 1795).

29. The biological uniqueness of humanity, discussed in Chapters 3 and 4, seems to provide no comfort.

30. Biologists refer to anatomical similarities which are not traceable to properties of a common ancestor as "convergent." The fusiform appearance of whales is convergent with that of fishes, although there was nothing fishlike in the whales' terrestrial, carnivorous ancestor.

31. This is illustrated in D. Lack, *Darwin's Finches* (Cambridge: Cambridge University Press, 1947).

32. Theodosius Dobzhansky, *Genetics of the Evolutionary Process* (New York: Columbia University Press, 1970), p. 131.

33. "Chance" events are those which could not be precisely predicted by existing scientific theories. This may be a deep property of the world, as, for example, the Heisenberg Uncertainty Principle, or it may be due to the coarseness or imprecision of our knowledge, as in some aspects of meteorological prediction. The appearance of chance may also arise from an unwillingness, or lack of urgency, in excessively complicating a model or a theory. For example, theories of population growth include death rates but do not attempt to predict exactly which organism will die on any given day.

34. D. P. Todes, *Darwin without Malthus: The Struggle for Existence in*

Russian Evolutionary Thought (New York: Oxford University Press, 1989). Also M. Nitecki, ed., *Ethics and Evolution* (Chicago: University of Chicago Press, in press).

35. London: Penguin, 1986.

9. Virtue and the Simple Life

1. Simon Schama, *The Embarrassment of Riches* (New York: Alfred A. Knopf, 1987); David Shi, *The Simple Life: Plain Living and High Thinking in American Culture* (New York: Oxford University Press, 1985).

2. When cars hit deer, or the converse, it is a kind of even fight, but there are other problems. Small mammals and reptiles have difficulty crossing any roads, particularly when there are barriers between the lanes. British highway designers put little open arches for hedgehogs in their dividers, but we have neither arches nor hedgehogs.

3. Stephen Bayley, *Sex, Drink and Fast Cars* (New York: Pantheon Books, 1986), p. 25.

4. It was the myth of simple preliterate society that let Huizinga focus only on "high culture." See Chapter 3.

5. T. Veblen, *The Theory of the Leisure Class*, rev. ed. (1912; New York: Modern Library, 1934).

6. Fred Hiatt and Margaret Shapiro, "Sudden riches creating conflict and self doubt: nation is pressed to define its values," *Washington Post*. February 11, 1990, p. 1. For the sake of those who may one day read very old copies of this book, I add that in 1990 $350 was a low but adequate weekly wage for a single person in the United States or Japan.

7. J. Huizinga, *The Waning of the Middle Ages* (New York: Doubleday, 1954), pp. 231–232. Also see Marston Bates, *Gluttons and Libertines: Human Problems of Being Natural* (New York: Random House, 1968).

8. Shi, *Simple Life*, p. 5.

9. K. R. Greenfield, "Sumptuary Law in Nürnberg," *Johns Hopkins University Studies in Historical and Political Science* series 36, no. 2 (1918): 1–135, is a fascinating detailed study of several hundred years of sumptuary laws in a single town. Shi and Schama both have some material on this; also J. M. Vincent, "Sumptuary legislation," in *Encyclopedia of the Social Sciences* (New York: Macmillan Co., 1948), pp. 464–466.

10. See note 1.

11. There were similar purple dyes produced from snails in other parts of the world, most notably the Yucatan peninsula, but discussion of these, as well as the detailed discussion of the biology and chemistry and precise religious roles of purple, would be too large a subject for this chapter. A recent symposium volume, E. Spanier, ed., *The Royal Purple and the Biblical Blue: Argaman and Tekhelet* (Jerusalem: Keter, 1988), and a small book by Lloyd Jensen and Frances Jensen, *The Story of Royal Purple* (Champaign: Garrard Press, 1965), provide an entree into the biology, chemistry, and social history of the color purple.

12. This is an informal estimate provided by David Slobodkin, M.D., from his experience in American hospital emergency rooms. It includes the consequences of drunken arguments and automobile accidents attributable to drinking as well as emphysema, burning mattresses, and cirrhosis.

13. Alexander Petrunkevitch, an arachnologist who was August Weismann's last doctoral student and one of my teachers, told me, sometime in 1947, that he was requested to attest to his belief in "a personal Devil" before he could be appointed to a position in the Yale biology program in 1911. I can't recall whether or not he assented to it. When I knew him he was a most devout atheist. It may have been a routine form, but it is interesting that it still existed as a question at all.

14. W. R. Cross, *The Burned Over District: The Social and Intellectual History of Enthusiastic Religion in Western New York, 1800–1850* (New York: Farrar, Straus & Giroux, 1950), p. 198, quoting *The Baptist Register* (Utica), June 23, 1826, p. 67.

15. Perhaps it was to supply an empirical basis for this supposed intrinsic property of humanity that the theories of human sociobiology were developed. Had they been valid, they would have verified the insights of the early Protestant divines as to the source of ethics, while demonstrating that science could maintain ethical sources similar in content to those claimed by the fundamentalists, but without requiring the supernatural.

16. L. Foster, *Religion and Sexuality: The Shakers, the Mormons, and the Oneida Community* (Urbana: University of Illinois Press, 1984), provides a sympathetic history from the beginning of the Shakers until the end of the nineteenth century. M. Marshall, *A Portrait of Shakerism: Their Character and Conduct from the First Appearance of Ann Lee in New England, Down to the Present Time. And Certified by Many Respectable Authorities.* Printed for the Author 1822 (Facsimile, New York: AMS Press, 1972), presents an impressive set of strongly negative affidavits from dissidents.

17. W. M. Miller, Jr., *A Canticle for Leibowitz* (New York: Bantam Books 1955), p. 36.

10. Masters of Reality

1. S. Johnson, *A Dictionary of the English Language* (London: W. Strahan, 1755). W. A. Neilson, ed., *Webster's New International Dictionary of the English Language,* 2nd ed. (Springfield: G. and C. Merriam, 1947). Dr. Neilson had honorary degrees from both Yale and Harvard and earned master's and doctor's degrees from Harvard, but did his undergraduate work in the University of Edinburgh.
2. Funk and Wagnall's *Dictionary.* I used an old copy, whose front pages had worn out long ago. It is lodged in the hall of the Woodrow Wilson Center in the Smithsonian Institution Castle, if the precise reference really matters. The attribution of a specific curative power resident in each plant species goes back at least to Dioscorides, who made the wonderful simplistic assumption that, since nature is basically benign, at least if properly understood, not only does each plant have a curative power but it carries a "sign" of that power in its anatomy. For example, the herb Hepatica has lobulate leaves so that it must help in liver disease, and the reddish spot on the flower of foxglove signs it as having importance for heart function. In fact foxglove does provide the drug digitalis, which affects the heart.
3. *New Catholic Encyclopedia,* vol. 13 (New York: McGraw Hill, 1967), p. 230. This is part of entry "Simplicity of God." Simplicity has no particular entry of its own apart from God.
4. This need not be true. It has recently been documented for the history of medicine in Germany and Austria, from 1870 to 1950, that many doctors had as their personal goals the health of the "nation and race" rather than that of the individual patient. See P. Weindling, *Health, Race and German Politics between National Unification and Nazism 1870–1945* (Cambridge: Cambridge University Press, 1989).
5. This is critical in the distinction between tractable and intractable science. See L. Slobodkin, "Intellectual problems of applied ecology," *Bioscience* 35 (1988): 337–342.
6. "This world" includes all of the universe which can by any instrumentation whatsoever be caused to produce sensory information, either by direct observation, instrumental amplification, or recording. It also includes events that could have done so had observers been there but which now must be inferred from consequences connected to them by theories. This is too small a playing field for mathematicians.

7. W. V. Quine states the same idea differently: "The truths of pure mathematics and logic . . . [are] true in all possible worlds." *Philosophy of Logic* (Englewood Cliffs: Prentice Hall, 1970), p. 4. That is, the facts of this world don't matter for the truth of pure mathematics or logic.

8. In the past couple of centuries there has been a distinction made between "applied" and "pure" mathematics. Applied mathematicians deal with problems that arise out of scientific, technological, or administrative concerns. The distinction is not really critical for my purpose.

9. A relevant anecdote was told by Sir Peter Medawar, who gave a copy of one of his books, replete with mathematical equations, to Waddington, an excellent philosophically inclined embryologist with almost no sense for mathematics. Waddington told Medawar that he enjoyed his book immensely. Medawar then asked "How could you? You can't understand mathematics." Waddington replied: "I hummed those parts."

10. A summary of the hydra experiments is given in L. Slobodkin, K. Dunn, and P. Bossert, "Evolutionary constraints and symbiosis in Hydra," in P. Calow, ed., *Physiological Ecology* (Cambridge: Cambridge University Press, 1986), pp. 151–167. Further developments of the theory, with a discussion of the dangers in the approach, are in M. Gatto, C. Matessi, and L. Slobodkin, "A physiological approach to ecology and evolution of simple organisms," *Evolutionary Ecology* 3 (1989): 1–30. The general problem of determining dimensionality of a cloud of points is treated in textbooks of statistics, and particularly studies of principal components analysis.

11. It is extremely unlikely that anyone actually thinks in a geometric way about a five-dimensional space, and a great deal of four-dimensional visualization is more poetic than scientific. W. Hurewicz and H. Wallman, *Dimension Theory*, rev. ed. (Princeton: Princeton University Press, 1948), is a standard elegant treatment of the mathematical idea of dimensions.

12. The tension in this field has been discussed in detail by David Hull, *The Evolution of Science* (Chicago: University of Chicago Press, 1988). "Parsimony," one aspect of simplicity, figures prominently in these arguments, but E. Sober suggests that much of the argument is irrelevant to the purpose of the classification and that this is an area in which process has displaced the focal role of purpose. See *Reconstructing the Past: Parsimony, Evolution and Inference* (Cambridge: MIT Press, 1988).

13. I am assuming that the identity of a real number is independent of the number of digits or other symbols with which it is expressed, so that 5, 4+1, square root of 25, and 10/2 are all reducible to the same single simple number. Complex numbers do not reduce to one number. They require at least two, each in its own dimension.

14. This carries the obvious danger that parts can be arbitrarily united or distinguished. Is "A and B and C" one statement or three? This is sometimes important but in our limited perspective I let it pass.

15. It doesn't matter whether or not you know what a matrix is, except to understand that its inversion is sometimes a good thing to do and takes an awful lot of arithmetic steps. If you are interested in the problem I was working on refer to L. B. Slobodkin, "Energy in animal ecology," *Advances in Ecology* 1 (1962): 69–103.

16. A recent text is R. Sommerhalder and S. C. Westrhenen, *The Theory of Computability: Programs, Machines, Effectiveness and Feasibility* (New York: Addison-Wesley, 1988). Chapters 9–12 are particularly relevant, but the area is quite technical. A brief nontechnical introduction is Hans Bremermann, "Complexity and transcomputability," in R. Duncan and M. Weston-Smith, eds., *The Encyclopedia of Ignorance* (New York: Pergamon Press, 1977), pp. 167–174.

17. This terminology is derived from tomato plants. An indeterminate tomato plant continues to grow and bear fruit until it is killed by frost. Determinate plants have a finite limit on growth and then fruit and are finished.

18. J. Boswell, *The Life of Samuel Johnson*, vol. 2, p. 203. (New York: Oxford University Press, 1933). "Edwards (A classmate of Johnson's at Pembroke). 'You are a philosopher, Dr. Johnson. I have tried too in my time to be a philosopher; but I don't know how, cheerfulness was always breaking in.'—Mr Burke, Sir Joshua Reynolds, Mr. Courtenay, Mr. Malone, and, indeed, all the eminent men to whom I have mentioned this, have thought of it as an exquisite trait of character. The truth is, that philosophy, like religion, is too generally supposed to be hard and severe, at least so grave as to exclude all gaiety."

19. This is not at all connected with creating artificial languages for everyday street use, like Esperanto or various sign languages for the hearing impaired.

20. Elliot Sober, *Simplicity* (Oxford: Clarendon Press, 1975), is one of the best of these studies that I have found. Sober himself has recanted much of its contents for reasons related to the discussion in this chapter.

21. M. G. Bunge, *The Myth of Simplicity: Problems of Scientific Philosophy* (Englewood Cliffs: Prentice-Hall, 1963).

22. M. Hesse, *The Structure of Scientific Inference* (Berkeley: University of California Press, 1974), p. 225.

23. M. Adams, *William Ockham* (Notre Dame, Indiana: University of Notre Dame Press, 1987), p. 60. The quotations from Ockham are also from this work. She notes that Ockham had to be argued into grasping his razor; "Walter Chatton convinced [Ockham] that the objective existence theory violated the principle of parsimony, better now known as Ockham's Razor," (p. 102). "Objective existence seems to be the existence due to an act of intellection—the existence of what is being thought about." (p. 156.)

24. George Steiner, *Martin Heidigger* (New York: Viking, 1978), p. 150.

25. In Ann Arbor, Michigan, sometime around 1960, Martin Buber fell asleep in the back row of an auditorium while five scholars heatedly disputed points of his writings. When he was awakened to arbitrate, he slowly walked the length of the auditorium, climbed onto the stage and said (as well as I can remember): "Gentlemen, what are your doing? You will make it so that people will say that in the last half of the twentieth century there was a philosophy called 'Buberism.' That is not my intention. All I am doing is pointing."

26. A. Lamorisse, *The Red Balloon* (Garden City: Doubleday, 1956).

27. B. A. Huberman, "The adaptation of complex systems," in Goodwin and P. Saunders, eds., *A Review of Theoretical Biology: Epigenetic and Evolutionary Order from Complex Systems* (Edinburgh: University of Edinburgh Press, 1990), pp. 124–133.

28. René Thom, "An inventory of Waddington concepts," in Goodwin and Saunders, *Review,* pp. 1–8.

Closure

1. L. B. Slobodkin and P. Bossert, "Coelenterata or Cnidaria," chapter 5 in James Thorp and Alan Covich, eds., *Ecology and Systematics of North American Fresh Water Invertebrates* (New York: Academic Press, 1991).

2. In the sense of Thomas Browne's *Pseudodoxia Epidemica.* Among what I consider to be the "vulgar errors" of current thinking about ecology and evolution are the simplistic equations between single properties and evolutionary fitness (for example, the equation of fitness with litter size, longevity, total lifetime reproductive rate, or physiological vigor). Eventually these will vanish, like the curious test of the success of testicular gland implants and injections supposedly used by the physiologist Claude Bernard more than a hundred years ago. He measured

his post-implant virility by how far he could urinate, standing with his hands behind his back.

3. I follow the procedure outlined by Veblen in the Preface to his *Theory of the Leisure Class*: "Partly for reasons of convenience, and partly because there is less chance of misapprehending the sense of phenomena that are familiar to all . . . the data employed to . . . illustrate the argument have by preference been drawn from everyday life, by direct observation or by common notoriety, rather than from recondite sources at a further remove. It is hoped that no one will find his sense of literary or scientific fitness offended by this recourse to homely facts."

4. The current publicity for the "Gaia hypothesis," which seems to impute a kind of teleological, holistic "wisdom" to earth, is a relevant example. If the earth is believed to be a kind of benevolent Shinto-style mother goddess, it will not help us in practical investigations and may substitute a simplistic, wild, new idolatry for science.

5. John Eccles, *Evolution of the Brain: Creation of the Self* (London: Routledge, 1989), chapter 4.

6. The meaning of "healthy," in this context, hinges on what is meant by "considered."

7. C. Taylor presented an excellent analysis of the operational aspects of "self" (Chapter 1) but then introduced an appeal to intuition to avoid a relativistic position on morality.

8. I have made several "advances" of that type. For example, I demonstrated that considerations of size and age were actually important in the growth of populations of animals, and could generate oscillations even in a constant environment; see "Population dynamics in *Daphnia obtusa* Kurz," *Ecological Monographs* 24 (1954): 69–88. This study not only demonstrated that the "sigmoid" or "logistic" curve of population growth was irrelevant to metazoans but also that recondite cyclical events in the environment were not universally relevant to cyclic fluctuations in population abundance. I also demonstrated that plankton blooms that could not possibly be happening in normal sea water were actually occurring in abnormal sea water; see "A possible initial condition for red tides on the coast of Florida," *Sears Foundation Journal of Marine Research,* 12 (1952): 148–155. This was updated in "The null case of the paradox of the plankton," in E. M. Cosper, V. M. Bricelj, and E. J. Carpenter, eds., *Novel Phytoplankton Blooms* (Munich: Springer-Verlag, 1990), pp. 341–348.

9. Occasionally fantastic constructs can be developed by attempting to translate between words and pictures or the reverse. The reconstruc-

tions of mythology by Robert Graves, *The Greek Myths* (London: Penguin, 1960), were based on his assumption that the verbal traditions of the Greek myths were based on attempts to tell stories about pictures. He then translated back from the stories to what he imagined must have been in the pictures, and then retranslated from the purely hypothetical pictures back into new prose narrative. These charming literary creations and the process by which they were produced must be considered as a satire of normal empirically oriented scholarship, almost in a class with the work of a Tasmanian entomologist who, I was assured by Evelyn Hutchinson, based his published interpretations of carboniferous insect fossils on information from a spiritualist medium named Marjorie whose control had been to the carboniferous.

10. For example, simple organisms may contain complex cells, but this requires careful contextual definitions.

11. This still stops short of condoning major plagiarism. Also, getting caught at falsifying data is usually the end of a career or a reputation. But even here there are exceptions, particularly if your conclusions were basically correct. For example, Mendel's genetic data are generally conceded to have been entirely too good. It had been polished either by Mendel or his gardener.

12. Michael Schrage, "Media-hype science often unrealistic," which appeared originally in the *Los Angeles Times* but which I read in the *Times of Tokyo,* December 24, 1989.

Acknowledgments

This project began when Stephen J. Gould suggested to Howard Boyer that it would be nice if I were to enlarge on my 1986 Presidential Address to the American Society of Naturalists. Howard persuaded me to try it and continued to encourage me throughout the writing process. Both of us were surprised with how it developed.

My Stony Brook colleagues Lev Ginzburg, Doug Futuyma, Pat Bossert, Ken Dunn, and others colored my thinking. However, the writing mainly occurred in quiet places far from home, while on sabbatical leave. I gratefully acknowledge the hospitable support of the Applied Mathematics Department of the Weizmann Institute in Rechovoth, Israel; the Istituto di Genetica et Evolutionistica and the Collegio Nuovo in Pavia, Italy; Tsukuba National University, Tsukuba, Japan; and the Woodrow Wilson International Center for Scholars in Washington, D.C.

At each place there were memorable discussions that helped clarify ideas. From Zvia Agur, Carlo Matessi, and Marino Gatto I learned something of mathematical attitudes; from Daphna Margolin I learned about the interaction between art and applied ecology; from Katke Scudo about the art of garden design. (As it happened, the manuscript took on a life of its own and insisted I postpone consideration of both ecology and garden design. I hope to return to these subjects.)

Professor Koichi Fujii gave me the opportunity to live in Japan under most favorable circumstances. Dr. Kinya Nishimura transmitted his enthusiasm for the shrines and temples of Kamakura. Dr. Yukihiko Toquenaga and Professor Humitake Seki furthered my education in things Japanese, despite my persistent ignorance of the language.

The Woodrow Wilson Center provided a charming office in the "Old Castle" of the Smithsonian Institution, excellent luncheon conversations, occasional sherry, and a collection of bright people willing to consider questions on a wonderful diversity of topics. Mike Lacey, James Livingston, and others listened, read, and criticized. The Center also provided student interns, Joseph M. Mehl and Julie Anzelmo, who trudged between my office and the Library of Congress and the libraries of the Natural History Museum, canvas book bags loaded with good things. Miss Anzelmo also did translations and hunted down quotations in exemplary fashion.

Early stages of the manuscript were read by Aldona Jonaitis, Roy Rappaport, Mike Rosenzweig, Dan Dykhuizen, Scott Ferson, G. Evelyn Hutchinson, and Dan Botkin. They were all extremely busy people, and I am grateful for their time. They suggested important improve ments. Finally, I want to thank Susan Wallace, Senior Editor at Har vard University Press. Aside from her help with the mechanics of turning manuscript into book, her comments and questions brought the whole thing into tune.

My wife, Tamara (who also read proof), my daughter Naomi, my daughter-in-law Orly, and my sons, David and Nathan, each provided emotional support and insights into their respective fields of expertise, which would otherwise have been closed to me. My grandsons showed me how very good, very new minds work. On rereading, I am impressed with how much of the book consists of material learned from my parents and grandparents.

I thank the Ontario Science Center for permission to reproduce a photograph of Rowland Emett's "Lunarcycle," Susan Bell of the Air and Space Museum, Washington, for the photograph itself, and Paul Terry for the text used in the caption. The photographs of the Thou sand-armed Kannon were taken by Asuka-en, Nara, Japan. Professor John Rosenfield of Harvard University very kindly troubled himself to help me locate these photographs, although we have never even met. The Jewish and German "mandalas" are photographs of objects in the collection of Dr. Ira Rezak of Oldfield, New York. The Tibetan mandala is rephotographed from *The Way to the Centre* by Willy H. Fischle (Malvern: Robinson and Watkins, 1978). The diagrams of DNA struc ture and the translation of DNA into protein are reprinted, with permission, from Karl Drlica, *Understanding DNA and Gene Cloning: A Guide for the Curious* (New York: John Wiley, 1984), figures 3-1 and 3-2. Copyright © 1984 by John Wiley and Sons, Inc. The lines from T. S. Eliot are from *Four Quartets* and are reprinted by permission of Faber and Faber, Ltd., London, and Harcourt Brace Jovanovich, Orlando, Florida.

Index